JETPACK DREAMS

JETPACK DREAMS

One Man's Up and Down
(But Mostly Down)
Search for the
Greatest Invention
That Never Was

MAC MONTANDON

DA CAPO PRESS
A Member of the Perseus Books Group

Library of Congress Cataloging-in-Publication Data
Montandon, Mac.
   Jetpack dreams : one man's up and down (but mostly down) search for the
greatest invention that never was / Mac Montandon.
      p.   cm.
   ISBN 978-0-306-81528-7 (alk. paper)
   1. Personal propulsion units.   2. Montandon, Mac—Travel.   I. Title.   II. Title:
Jet pack dreams.
   TL717.5.M66 2008
   629.133'3—dc22

                                                                    2008018405

First Da Capo Press edition 2008
Published by Da Capo Press
A Member of the Perseus Books Group
New York
www.dacapopress.com

Da Capo Press books are available at special discounts for bulk purchases in the
U.S. by corporations, institutions, and other organizations. For more information,
please contact the Special Markets Department at the Perseus Books Group, 2300
Chestnut Street, Suite 200, Philadelphia, PA 19103, or call (800) 810-4145,
extension 5000, or e-mail special.markets@perseusbooks.com.

10 9 8 7 6 5 4 3 2 1

For my brother, Asher

Up in the air, junior birdmen
Up in the air, upside down
Up in the air, junior birdmen
Keep your noses off the ground.

—From an old camp song

CONTENTS

JETPACK DREAMS

CHAPTER

1

The Voluptuous Panic, Part I

See my lips tremble and my eyeballs roll, Suck my last breath, and catch my flying soul.

—Alexander Pope, "Eloisa to Abelard"

"Ohboyohboyohboyohboy!"

Oh, boy? Since when do I exclaim, "Oh, boy" at moments of unusual excitement? It seems so quaintly nerdy. I mean, that's what my grandpa hollered the first time he took me on Space Mountain at Disneyworld.

But, as it turns out, this is exactly what I yelp as I'm being whisked off the ground by a 1,600-pound Cessna Skyhawk and suddenly cruising 9,000 feet above Belfast, Ireland, sharing the cramped, pungent cabin with a wispy-bearded nineteen-year-old theoretical physics major at the local Queens College, whom I've known for all of two hours and whose parents I've never met, though I've spoken to his father a couple of times on the telephone and I guess he seems nice enough.

Oh, boy? I can only hope there isn't a recording of my odd, panicked chirping out there, the better to blackmail me and thus derail any latent presidential yearnings that I might someday develop.

In my defense, I've never flown in a Cessna before, and it's only when Will turns to me and says with a vaguely sinister smirk, "So, how'd you like to experience zero gravity?" that it strikes me: I don't even care much for roller coasters, let alone tiny, wobbly, metal death boxes operated by more or less complete strangers. I remember that, not long ago, New York Yankees pitcher Corey Lidle crashed one of these babies into an apartment building on the East Side of Manhattan.

"Um, yeah, sounds great."

With a noise-canceling headset on, my voice sounds as if it's coming from deep inside me, so deep that it could be someone else talking— someone nasally congested, more than a little unnerved, and alarmingly close to throwing up all over the beautiful round instruments of the dashboard. (I'd like to blame the cream of parsnip soup, the lamb shoulder with pureed spinach and carrots entrée, and the glass of Shiraz, eaten about an hour earlier in a weirdly ornate Belfast restaurant, at which Will had thoughtfully made reservations. I'd like to blame lunch, but it's probably not lunch's fault.)

"Okay, then, you might feel a little unusual, but it'll only last a couple of seconds."

"Sure, sure." A couple of seconds—ha! What's a couple of seconds? This . . . is . . . them! That's all. I could take anything for a couple of seconds, right? A short skip on smoldering coals, the briefest of CIA interrogations, a single reaction shot from the Tara Reid cinematic oeuvre.

Will flips a switch, the plane drops as if we've been shot down, and my stomach leaves my body. I taste my teeth. My notebook floats up between our two seats, like the best bar mitzvah party magician's trick ever. Zero gravity is two seconds of glorious terror. A voluptuous panic.

When we level out and my stomach returns to my insides, lunch is angry, pissed really, inconsolable; it wants out of this relationship. It wants to roam freely among the cabin and not be tied down to just one man. I, on the other hand, am hoping we can still work something out. Sorry as this might sound, I will remain woozy and clammy to the touch deep into the night, until long after I've said good-bye to Will, staggered into a taxi, and boarded a train back to Dublin.

But for now I'm trying to appreciate the grandeur below us, which is, nausea be damned, magnificent. We aren't so much flying as floating—and below us I can see a gorgeous late-fall mosaic made from soft emerald hills, mauve-laced fog, and the choppy blue-gray sheet of Belfast Harbor. From our current vantage, it's hard to believe this land was once so inflamed. I turn to my left, and Will is grinning broadly, a flicker of danger in his eyes.

How did this happen? How did I end up cruising above Belfast with a teenager whose full name, Will Breaden-Madden, morphs so easily in my mind to "Will Breed Madness"? So far away from my wife, Catherine, our two young daughters (two-and-a-half-year-old Oona and four-month-old—four months!—Daphne), and our mostly very pleasant Brooklyn, New York, life on Prospect Park? Why would I—I, who am very much against pain, particularly my own—be doing this?

All very good questions. To try to get to the bottom of them, to give them the answers they deserve, we must go back in time a few months, back to the long, hot summer following my thirty-fifth birthday. It is there that several events—related, as I would later discover—sent me down the mysterious path leading to Will Breaden-Madden.

Prior to turning thirty-five, I'd never given much thought to age. Although I didn't go around broadcasting it, I was, I suspect, one of those annoying people unimpressed by the humbling limp toward death, and had all but tricked myself into believing all that malarkey about age being just a number, about how you're only as old as you feel, and so on. You know—the language of bran-cereal marketers and purveyors of erection medication. Despite being a devoutly unreligious person, I suppose I'd adopted an almost Zen Buddhist–like ability to live in the now, though I never would have called it that. If I'd called it anything, it probably would've been denial, one of the few skills I'd say I'm fairly expert at.

But then an odd thing happened: I began paying closer attention to the *New York Times*'s obituary pages. I'm not sure why I started doing that, but I distinctly remember reading those pages and—how to put this delicately?—*freaking the fuck out*. In my own quiet way. It wasn't one thing in particular that alarmed me but rather a collection of statistics. That is, I began to notice just how many people die in their seventies. Or even younger.

Take today, for instance. Let's see who we have: "Larry Sherry, 71, M.V.P. of 1959 World Series" and "Sheldon Fox, Architect and Manager, Dies at 76" and also "Larry Zox, 69, Abstract Painter of Dynamic, Geometric Works," then, finally, "Dennis Payton, 63, of the Dave Clark Five." Better make that the Dave Clark Four.

I'm especially drawn to Sherry's story. An unknown midseason call-up, the Los Angeles Dodgers relief pitcher was born with clubfeet. He was muddling along in a Venezuelan winter-ball league in 1958, but then developed a wicked slider and less than a year later was crowned the best baseball player in the galaxy. After earning the MVP award, *The Ed Sullivan Show* called, and Sherry realized he'd better go shopping—he didn't own a single suit.

My kind of guy—a nobody who, for one week at least, was better than anyone else, better than Sandy Koufax, Don Drysdale, Gil Hodges, and Duke Snider, all of whom were on his team. Still, that didn't stop him from dropping dead of cancer at an age exactly double what I am now.

For some stupid reason—looking at you on this one, science—I'd gotten it into my head that we humans were living longer, healthier, and happier lives. Of course, rationally, I know there are reams and reams of research pages indicating just that. But I guess I optimistically wanted to believe this meant *all* of us were living longer. That's not what the research was coming up with, and the *Times*'s obituaries were only too pleased to bear it out.

And so—math, an area, I must confess, in which I am lacking. Okay, I suck at it (a product of progressive schools, I'm afraid, with too many

too patient teachers and their quirky notions about learning, not giving grades, and, yes, math). Yet I could tackle this simple equation: 35
X 2. Gulp. You know what that equals? It equals holy crap!

So that's how it began: with a premature midlife crisis. (Let me just
say that the preceding sentence concludes with three of my least-
favorite words; put them all together and, well. . . .) From there, if one
isn't careful, one will soon be contemplating such unpleasant things as
lifetime achievements, legacy, and potential and the damnable fulfill-
ing of it.

Michelangelo, after all, was only thirty-three when he scored the
Sistine Chapel gig. Mozart wrapped up what many consider a pretty
decent career and played his final concerto by the time he turned
thirty-five. As of this writing, the peculiarly Dorian Gray–ish Beyoncé
Giselle Knowles is still only twenty-seven years old.

The unexamined life may not be worth living, but it sure is more
fun. (Plato's *Apology*, by the way, was written when our man was but a
sprightly twenty-eight.) And so it came to pass that, for the first time
in my life, I had trouble sleeping.

Yes, I'd done some things I was proud of. Through sheer force of will
I'd overcome a dismaying lack of connections to turn a flimsy hippie-
school bachelor's degree in modern literature into a career as a journalist. I'd written a few short but mildly charming stories for top
media companies, edited a book on musician's musician Tom Waits,
and worked on staff at a handful of national magazines. I'd had the
good sense to marry a woman not only kind, smart, and sexy but
also hailing from a ridiculously large Irish Catholic family (she's one
of thirteen siblings), thus increasing the number of roofs we could
theoretically crash under should things ever turn truly dire. At the
time, we had one beautiful, superspirited, hilariously extroverted
daughter, with a second well on her way. By many standards, I had
it made.

Then how to explain the nagging feeling that I was halfway home in a life not only unexamined but also unexceptional? The truth is, I desperately craved the exceptional, the big prize, the earth shattering. I wanted my Moby Dick. Just a single World Series MVP award. Really, who doesn't? And the thought that it might be out there waiting for me, while the yarn of my life uneventfully unspooled, was enough to keep me up nights.

Here, then: the *Cliff's Notes* of my mediocrity. As a teenager, I was a decent athlete; at one point, some coaches felt I showed Division III promise in both baseball and basketball. But I soon stopped growing (at five foot eight) and discovered, in this order, girls, cigarettes, and New Wave music. I left sports behind. Later, I dropped out of New York University after my freshman year, moved to Los Angeles, rented a Hollywood dump with an older, cooler cousin, bought a motorcycle, and took acting lessons. My teacher—an intense former character actor who had three names and an unruly thicket of dark hair—said he saw a lot of Spencer Tracey in me. I felt like a fraud and quit soon enough.

Next came San Francisco, where by default I became "lead singer" in a rock band that badly wanted to be the next Nirvana and may have had a shot at doing just that, except for the fact that its lead singer couldn't, you know, sing. After playing a few barroom shows, we tried to record a single, but the session fell apart when I wasn't able to nail the vocal track.

From there I landed back in the only place that would have me: the bohemian country club in the redwoods that is the University of California–Santa Cruz. There I was lucky enough to find a couple of inspiring lit professors and Catherine, to whom I've now been married for nearly seven years. Which brings us back to that summer of my thirty-fifth year.

It's funny where certain ideas come from. A few years after he invented the telephone, Alexander Graham Bell was called on to help locate a bullet lodged in the chest cavity of assassinated president James Garfield. The elaborate device Bell threw together found metal all right, but it was the metal bed frame Garfield was lying on. The bullet remained missing, but the metal detector was born. To this day beach bums the world over have a restless inventor and the unpopular twentieth president of the United States to thank for it.

The Wright brothers were just a pair of unassuming bike-shop retailers when they decided to change the world forever. It was while observing the motion of an empty bicycle-tube cardboard box that Wilbur Wright came up with his theory on wing-warping, thus unlocking one key to the challenge of flight. This allowed the wings to twist slightly, enabling the pilot to control an aircraft's direction without having to contort his own body to do so. Very important. By all accounts, Wilbur was less than elated to be working at the shop that day, or any other day, but if he hadn't been, who knows if you'd be going to Cancun next week.

The idea that ultimately took me to see Will in Belfast, and many other places, too, has a similarly modest origin. I was talking with my friend Jofie on the phone one day. Nothing especially noteworthy, just the usual business about sports, books, life, the future—those are the subjects we tend to stick to. But that day was different because at one point Jofie blurted out, "I mean, where's my fucking jetpack, you know?"

"Excuse me?"

"You heard me: Where's. My. Fucking. Jetpack? Look—you can hold your entire musical library in your pocket. We've got cell phones and laptops and the Segway—the fucking Segway—and you're telling me we can't have our own personal jetpacks by now? That's bullshit."

It should be noted that Jofie did not typically talk such a blue streak but swerved in that direction when he was especially fired up about something (or had been drinking). It should also be noted that I hadn't actually been telling Jofie that he couldn't have a jetpack. Even so, I could totally see where he was coming from and, more important, I could almost immediately see where this might go. The words were barely out of his mouth when they began revving up inside me, too: Where's his fucking jetpack? No—where's *my* fucking jetpack?

For weeks after we spoke, I couldn't get Jofie's question out of my head. Having come of age in the *Star Wars* era, I'm part of the post-moon-landing, post–World's Fairs generation, and thus was once very certain that by no later than the year 2000 we would most definitely be living *in the future*. And in that glorious future, we would have long ago traded in our dirt-streaked Hyundais and battered Kia Sportages for shiny metal backpacks with jet engines welded to them, the better to launch ourselves like angels into the atmosphere and soar over bumper-to-bumper rush-hour traffic. Or at the very least, we'd have hovercrafts and flying cars. But really, the future meant jetpacks.

That, after all, is what I'd been told since I was a very young boy. I was told that at the movies, where the single coolest character perhaps ever, George Lucas's bounty-hunting Boba Fett, blasted off at the drop of a Wookiee scalp with his clunky, dusty, and most righteous jetpack—the thing looked like a portable missile launcher dreamed up in the Renaissance—in the third installment of the original *Star Wars* trilogy. And I was told that when I snuck some TV (progressive schools!) and caught the *Jetsons*. And when I flipped through magazines at the dentist's office and came across a Canadian Club whiskey ad featuring a man with great teeth, a hot girlfriend—and a jetpack. And I was even told it in real life, when I'd press my dad or grandfather to tell me the truth, and they would say, "Yeah. Yeah, I could see that." Aha!

But as the *New York Observer* noted in early 2007, "Like kids outgrowing Santa Claus, we've spent the past seven miserable years learning to stop dreaming about the World of Tomorrow. Why would we? In the continued absence of solar-paneled jetpacks, plutonium-powered time machines or even fully electric (forget flying) cars, most of us still arrive at our still-earthbound offices via that great marvel of 1904, the subway. Which rarely gets faster, cleaner, cheaper or more frequent, but instead every day further erodes, like the ruins at Troy."

Oh, great. And it gets worse: "Americans have always assumed that one day we'd awaken in our utopian future, like tourists at Disney World wandering happily from Frontierland into Tomorrowland. We envisioned it in books, in movies, on TV, in bedtime stories. But we took the future for granted, as if it were a wife. And maybe it escaped this neglectful marriage, changed its name and skipped town."

No, please, don't skip town, future. C'mon! Disney's EPCOT Center opened for business when I was a very impressionable eleven years old, so of course I still believe in you, future. EPCOT—Experimental Prototype Community of Tomorrow. Tomorrow!

At the same time, I also felt as if Jofie's question had plucked some deeper psychological string inside me. We live in a world of caterpillars dreaming of butterflies. From Icarus to the Wright brothers, from Pan Am to Superman, and da Vinci to Delta, the thought of flying is the most enduring of human daydreams. And night dreams, too: when we each lie down to sleep, there is a very good chance our real dreams will be crammed with heart-tenderizing images of majestic, effortless flight. The individual desire to fly—not as a group in the frustrating, frightening settlement of an airplane but as a comic-book hero might, as a machine of one—is an essential aspect of human consciousness.

And it starts early. Here's a recent scene from Chez Montandon. I'm on the couch reading or contemplating my shortcomings or something; Oona toddles by. She takes no real interest in me until I look up and ask, "Hey, Oon, how about an airplane ride?" She lights up like a runway.

Soon I'm flat on my back on the living room rug with my daughter trapped between my calves, and she's above me, soaring, pretending to fly. She has her arms out in front of her in a perfect Superman pose, though I'm pretty sure she's never seen Superman. All fourteen of her teeth are exposed in a wide grin. Every one of the twenty-nine pounds of her is tense with delight. She is, in a word, ecstatic. From my perspective, it's not much of a stretch to imagine her with a tiny made-for-tots jetpack strapped to her back.

Dreams of flying hit me young, too. The forest-green house in which I grew up in a Baltimore suburb had broom-riding witches carved into its black shutters. Knowing they were just outside my bedroom window scared and thrilled me. The witches were designed with brooms pointing upward, forever frozen at takeoff.

A favorite book in our household library was a Russian folktale called *The Fool of the World and the Flying Ship.* My mom tirelessly read and reread the colorfully illustrated paperback, which tells the story of a simpleton peasant competing against his two more clever brothers for the privilege of marrying the Czar's daughter. The winner of this competition will be the first man who can deliver to the Czar—what else?—a flying ship, which I like to think of as an early jetpack prototype.

And so the titular Fool of the World sets off to find the Czar the object of his desire. He soon encounters an old man, with whom he kindly agrees to share his modest food and water. This being a folktale for children, when the Fool opens his parcel he finds his stale bread converted magically into a feast, his water miraculously distilled into brandy.

The Fool and the old man eat and drink until the Fool passes out drunk. When he wakes, the old man is nowhere in sight but—score!—in his place is left a handsomely carved wooden flying ship that allows the Fool and the Czar to sail the sky. Soon the Fool and the Czar's daughter are happily married.

As a four year old the message could not have been clearer: be kind, get sauced, and a jetpack will soon be yours. (Note: To date this formula has not produced the desired effect.) As a boy I didn't give a damn if my GI Joe came with Kung Fu grip; my sights were set higher, on his One-Man Jetpack Assault Module.

These relentless flying fantasies come from all over. They come from religion—Adam and Eve, after all, didn't bicycle from grace or swim from grace, they fell, and had they had jet engines strapped to their backs everything might've been different. They come from science— Leonardo da Vinci drew detailed sketches for wooden wings to be worn by a human pilot more than 500 years ago. A note in the margin of one da Vinci drawing reads: "The Great Bird Will Take Flight." The fantasies are handed down to us, also, from mythology—who hasn't heard the cautionary tale of Icarus, his wax wings, and what happens when you soar too near the sun? (This myth continues to resonate loudly with contemporary rock bands: U2, Megadeath, Rush, Kansas, Faith No More, and Iron Maiden have all referenced Icarus in song at one time or another. So there's that.)

Today there are countless ways to launch oneself into the sky, from trampolines and pogo sticks to hot-air balloons, helicopters, blimps, light airplanes, and JetBlue to more eccentric methods like becoming a human cannonball, bungee jumping, hang gliding, paragliding, para-chuting, donning a wing suit and making like a bird, or even, in a cou-ple of insane, mostly Dutch, examples, affixing rocket engines to one's ski boots in order to streak through the air after being dropped from a plane. Oh, and if you dare, you can also do like the ballsy character George Plimpton once wrote about named Larry Walters, who in 1982 attached forty-two hefty helium-stuffed weather balloons to a Sears, Roebuck lawn chair and climbed to nearly 17,000 feet above the Cali-fornia coast.

Flying fantasies confront us at every turn. And there is evidence it's been this way for a very long time. A recent fossil discovery revealed that the first mammals capable of gliding flight lived many millions of years ago. The fossil in question belongs to a Chinese squirrel-like creature, which possessed a stretchy membrane between its front and back legs that served as wings. Some scientists believe the animal may have lived as long as 164 million years ago, meaning that mammals were taking to the air before birds.

Yet still, we, the ultimate mammals, have no jetpack.

Flash forward 164 million years. By the summer of my thirty-fifth year, my life was evidently half over, and I'd come to accept that I was never going to play shortstop for the Baltimore Orioles or be the next Spencer Tracey or Kurt Cobain. That's when the question was zapped my way like a laser shot from robot eyes: *Where's my jetpack?* Whatever happened to what must surely be the greatest promise never kept?

Soon an idea began to take shape. I could go out into the world, wherever it made sense to go, and some places that perhaps it did not, and find out what happened to our jetpacks. I mean, is this the future or is it not? And as a serious bonus, perhaps my quest would lead me to someone who could still make the dream come true. Some might not consider it on par with the stuff of Michelangelo or Mozart, but it was something I thought I could do.

As I began telling friends and colleagues about my plan, I quickly realized that Jofie and I were far from alone. In fact, just about every (male) friend I told, regardless of age, responded with what can only be called spazzy enthusiasm. One guy, a majestically ironic twentysomething magazine editor, confessed maniacally to me: "Yes! Yes, totally. I remember sitting in class when I was a little kid just wishing I could blast off with a jetpack and get out of there." He gripped the side of his chair to demonstrate and shook like a hipster epileptic while making throaty blast-off noises: "Shhhhrrrrrruugghhhh!"

We were having lunch in one of those Manhattan restaurants where the waitstaff consists entirely of supermodels, and by acting so publicly geeky, he was ensuring that he would never get to date one. By extension, my friend was basically telling me that he'd rather talk about jetpacks than have sex with a supermodel.

Most other people had similarly impassioned responses to the mere mention of the word—*jetpack*. But not always. Sometimes when I explained my plan to a female friend, her expression would say to me, "I have absolutely no idea what you are talking about, and in fact if a telemarketer called my cell phone right now, I'd answer." But that was rare, and at those moments all I had to do was pantomime wearing a backpack equipped with launch-capable, powerfully thrusting engines and taking off with it to win over such a friend. "Oh," she'd say then. "That does sound pretty cool."

You're damn right it does. Before I knew it, I'd become completely enchanted with the idea of hunting one down. Soon I started to feel all *Star Wars*–y, as more and more it began to seem like my destiny to bring the good people of the world their rightful jetpacks. In order to do so, I thought, I must first look to the past, to prepare to blast into the future.

CHAPTER 2

The Past Is Prologue?

I want to fly like an eagle.

**—The Steve Miller Band,
"Fly Like an Eagle"**

The jetpack was first popularized in a novella published in the August 1928 issue of a once widely read pulp magazine called *Amazing Stories*. The first science fiction magazine in the United States, *Amazing Stories* began in 1926, with the enticing tagline: "Extravagant Fiction Today: Cold Fact Tomorrow."

In "Armageddon 2419 A.D," author Philip Francis Nowlan used the time-honored narrative device known as the Radioactive Experiment Disaster Technique to transport his protagonist, veteran World War I pilot Captain Anthony Rogers, five hundred years into the future. In Tony's future, the planet is populated by people who concoct synthetic foods, wield ray guns, and have the good sense to travel via jetpack rather than automobile. (Hard-core fans will point out, however, that in early episodes Rogers's compact backpack contains a chunk of antigravity, making flight possible. This technology eventually morphed into rocket engines, in part, some have speculated, because readers wanted to see the machine at work in greater detail.) Our hero was soon blasting off the page and into the hearts of adolescent boys everywhere.

Interestingly, the cover of the *Amazing Stories* featuring Anthony's debut was devoted not to Rogers but to a story written in 1918 by a former chemical engineer named Edward Elmer Smith, "The Skylark of Space." The cover illustration shows a young man wearing what appears to be an old-fashioned leather football helmet, a form-fitting racing suit, and knee-high boots of the sort that remain in fashion with chic metropolitan women the world over. This is the future as imagined in post–World War I America. The man is hovering in midair with a very elaborate device strapped around his groin area, continuing over his shoulders and halfway down his back, not unlike a high-tech man-purse. With his right hand he is waving to an attractive, slightly neurotic-looking woman who is waving back to the *Skylark* from what must be the outskirts of her family's farm. In the man's left hand he is holding a joysticklike apparatus that glows green, indicating the groin-back machine is indeed on. This is either a very early depiction of an antigravity machine in action or the Skylark of Space is an incredibly talented, if perhaps a bit too flamboyant, leaper.

Nowlan's Anthony Rogers narrative proved such a hit with readers that it led in quick succession to a sequel, a wildly popular comic strip—the first science fiction strip on record—movie, and television franchises, and eventually, many years later, to video-game spin-offs. Despite the many iterations, the character is still most famously associated with the long-running comic strip, which debuted in 1929, wherein Rogers's creator smartly changed his name from Anthony to the much more rugged-sounding "Buck." That series, *Buck Rogers in the 25th Century*, ran regularly until 1967. At its peak it was syndicated to more than four hundred newspapers across the globe and translated into eighteen languages. According to legend, the character was so beloved that a Virginia department store celebrated Christmas in 1934 not with an appearance by Santa Claus but with an actor dressed as Buck.

For several years following the publication of Nowlan's tale, it seemed as if the technology that had so enchanted readers would remain solely in the realm of science fiction. Though the stubborn, frustrated genius Robert Goddard (much more on him later) had launched the first liquid-fueled rocket a good two years before Buck made his debut, for some reason it hadn't yet occurred to anyone to strap a real rocket to a real man's back and launch him heavenward.

In the early 1930s, however, that began to change. In Germany a very determined, fearless, and somewhat asinine young fellow thought to set off a few rockets while strapped to roller skates. A crowd gathered one day to watch his demonstration, and, as it turned out, they were in for a treat: after igniting the first blast, the poor guy was knocked off his skates and sent skidding across the ground for several feet until he came to a grinding stop—bloodied, burned, yet somehow still alive. Luckily for him, a cameraman caught the entire thing on film, and soon his folly was packaged as part of a reel of shorts shown in cinemas all over the world.

Fortunately, the hapless chap was not the only one in the world, or even in Europe, attempting such stunts. Around this time, a Russian fellow identifying himself as A. Andreev filed a patent for an oxygen- and methane-fueled flying device that could be worn on the back, with roughly three-foot wings extending to either side of the hopeful pilot. The intrepid airmen could crank a handle, which in turn would kick-start a windup spring motor and lift him off the ground. The patent featured a crude depiction of the apparatus, but there exists no record that the machine was ever even tested. Progress, or so it seemed, was on the march. "This is the first device of its kind that had any engineering detail at all," says Mark Wells, a research engineer for the Center for Applied Optics at the University of Alabama in Huntsville who has worked for NASA, the U.S. military, and corporate clients throughout his twenty-three-year career.

Nonetheless, jetpack development hit a barren patch after Andreev's patent. By now, the world had been seduced by the considerable

achievements of a pair of Ohio brothers and was in thrall to a different sort of flight—that of the airplane.

But the Pandora's box of jetpack fantasies had been opened and
absorbed into the collective consciousness of a generation. With the
Buck Rogers comic strip a smash in the 1930s, Hollywood caught up
with the phenomenon only toward the end of the following decade,
when the 1949 Republic Pictures film serial *King of the Rocketmen* debuted in theaters. The narrative wasn't much to behold, revolving
around a badass scientist named Jeff King and his attempts to undo
the evil criminal Dr. Vulcan, who had stolen an incredibly powerful
new weapon called the Decimator. Despite the hackneyed plot, the
film represents an important moment in the often sputtering history
of the jetpack, the first moving pictures, either real or imagined, of a
man lifting himself, without wings, into thin air.

And not always gracefully. King's suit consisted of a frowning
bucket for a helmet, a black leather jacket, a small backpack, and controls welded to a chest plate. While watching him flying—or wobbling
along often hidden safety wires—his arms outstretched à la Superman, one notes a disheartening lack of flame blasting from his twin
rockets. He doesn't so much land as hurl himself at, ideally, bad guys,
but far more often the ground. Perhaps this isn't surprising when one
learns that the stunts, such as they were, relied overwhelmingly on the
use of a trampoline. An inauspicious beginning to such visionary activity, but a beginning just the same.

King of the Rocketmen's twelve installments opened the eyes of Republic executives to the growing demand for science fiction matinees.
To reach this new audience, the company cranked out another serial,
1952's *Radar Men from the Moon*, featuring a character who would en-
dure for many years as a favorite among space-obsessed boomers—
Commando Cody. A group called the Nostalgia League offers a colorful
and succinct online synopsis of the Commando series:

George Wallace . . . appeared as Cody. His nemesis was Retik, evil ruler of the Moon, portrayed by stock studio villain Roy Barcroft. Unable to survive in the thin, dry atmosphere of the Moon, Retik sends an agent, Krog, to earth in an attempt to break down global defenses prior to an impending lunar invasion. Aided by n'er do well Graber (Clayton "Lone Ranger" Moore) and his partner Daly, Krog attempts to systematically destroy key targets using a truck mounted ray gun. To counter these deadly attacks, the government enlists the services of Cody Laboratories. In addition to his flying suit, the intrepid Commando utilizes a newly developed rocket ship to combat enemy forces locally and on the Moon. Retik's campaign is finally derailed in chapter twelve and he himself destroyed by his own ray gun during a failed return to his home planet.

And there you have it. *Radar Men from the Moon* incorporated the same airborne footage as *King of the Rocketmen*, but moviegoers didn't care, so long as they could get their flying fix. Republic Pictures subsequently dusted off Jeff King's rocket suit two more times, for the wonderfully titled serials *Commando Cody, Sky Marshal of the Universe* and *Zombies of the Stratosphere*. In 1955 the Cody adventures were spun off into a television series for NBC, thus allowing stargazers the country over to tune in on Saturday afternoon and, for a half hour at least, witness one of their greatest remaining unfulfilled dreams.

Meanwhile, the way things were going back in real life, it was not crazy to think that an actual, honest-to-goodness working jetpack might be available not too far in the future. In 1950 the world-class German scientist Dr. Wernher Von Braun, who years later would lead the Marshall Space Flight Center and help develop NASA's space-age-ushering Saturn V rocket, relocated from Fort Bliss to Huntsville to work for the burgeoning Army Ballistic Missile Agency (ABMA). Von Braun had a controversial legacy—in Germany he'd been a member of the Nazi Party and had built the first ballistic missile, the V-2, which was fired at British and Belgian targets near the end of World War II.

Like so many others before and after him, Dr. Von Braun reinvented himself after immigrating to the United States in the mid-'40s, turning his back on a violent past to look instead into what must have seemed a limitless future. By the time he landed in Alabama, the doctor was determined to explore new frontiers beyond the edge of Earth.

Von Braun took several of his Fort Bliss staffers with him to Huntsville. Among those he moved with was a young radar technician fresh from the laboratories of MIT named Thomas Moore.

Moore was one of a handful of men who formed Von Braun's inner circle; future Kennedy Space Program chief Kurt Debus was perhaps the best-known member of the group. Over the years, trailing in his boss's charismatic wake, Moore came in contact with some of the most famous flyers in the history of aviation, including Orville Wright and Charles Lindbergh, the latter of whom became an occasional Ping-Pong partner.

Tall and slim and always impeccably dressed in coat and tie, Tom Moore was not married, had no children, and seemed to have few interests outside of work. He was, however, a rabid reader of science fiction literature and a longtime fan of Buck Rogers. Prior to moving to Huntsville, in fact, he had often fantasized about what it would be like to strap a pair of rockets to his back and blast off, just like his comic-strip hero.

Mark Wells, the Alabama engineer, befriended Moore during the last few years of his life. He and his wife spent nearly every Sunday afternoon visiting with Moore at his home, sitting in his living room reminiscing, while perhaps also playing the part of surrogate son and daughter-in-law. He shared the older man's obsession with space exploration—Wells himself possesses an expansive library dedicated to the subject; on his office wall hangs a poster of Buck Rogers.

According to Wells, sometime soon after going to work for the ABMA program in 1950, Moore stumbled across the Russian Andreev's patent in a future-gazing book titled *Journal of Interplanetary Communications.* To call this moment an epiphany somehow doesn't do it justice. From that moment until he died in 1994 at the age of eighty-three, Moore obsessively, occasionally maniacally, pursued the singular solo flight machine we've come to know as the jetpack.

He made far more progress than anyone who had come before him. In 1951, with the backing of Von Braun and Debus, Moore secured a grant of twenty-five thousand dollars from the United States Army, which was interested in innovative ways to transport troops.

Building on the V-2's fuel system, a combination of ethyl alcohol, liquid oxygen, and hydrogen peroxide, and the English Stentor motor, which utilized a silver-based catalyst chamber, Tom Moore put together a device whose basic technology is still emulated by would-be rocketeers today. The idea was to have the highly reactive hydrogen peroxide pumped at high pressure into a Stentor-like (except monopropellant) minimotor where it would be superheated to about 1,300 degrees Fahrenheit, expand to roughly 5,000 times its size, and exit a pair of stainless-steel nozzles at hydrogen peroxide's characteristic velocity, or 3,000 feet per second. This, in turn, would yield some 300-odd pounds of thrust, enough to get a slender pilot off the ground. (A 100-mph major-league fastball whipped at an unfortunate batter is traveling at but a paltry 147 feet per second; compared to Moore's motor, the baseball is practically standing still.)

The inventor called his glorious apparatus a Jet Vest, which, admittedly, sounds more like an article of clothing you'd wear to prom to help seal the deal than a cutting-edge piece of hardware ripped from science fiction fantasyland. In Moore's sketches, though, the Jet Vest looked decidedly slick. The helmeted, flight-suited sky commander was adorned with a metallic fuel tank covering most of the length of his back. Where Andreev had thick and long wings, Moore envisaged stainless-steel piping jutting just far enough from the tank to allow the pilot to control pitch and yaw—the rotations around the side-to-side and vertical axes of an aircraft, respectively—with a pair of magnificent rockets. Drawings optimistically depict Jet Vesting soldiers fluttering up near a helicopter and effortlessly hopping a steep embankment. The Jet Vest was certainly a bulkier piece of machinery than Buck Rogers's svelte system, but not by much.

Too bad it never flew. Mark Wells says Moore told him he tested his Jet Vest on a safety wire, but there are no photos of such exercises and none of the existing hardware has ever been found. (There are pictures of Moore flying a Jet Vest prototype, but only on an elaborate and grounded harness using compressed air to generate lift.) Perhaps Moore destroyed it. After Moore threw every bit of engineering acumen he had into the project, the military deemed it "too outlandish." Once the funding ran dry, the government never offered additional money for research. "Even Von Braun went to bat for him," Wells told me. "But they couldn't get any interest."

Moore was crushed; it would remain his greatest professional disappointment. Still, he refused to throw in the lathe. He continued to tinker with his creation, exploring the use of jet engines at one point, and later a Lockwood wing to increase airflow and amp up the thrust. In the end, it proved to be only so much sound with the odd bit of fury. As Wells notes sadly, "Moore was very much involved with this stuff, though he didn't accomplish much after the 1950s." The military, meanwhile, had moved on, taking with it any dreams of a jetpack in the offing. The war in Korea was winding down just as the cold war was heating up—there was little time for the eccentric or quixotic.

It is a testament to the power a jetpack can wield on one's imagination that the story doesn't end here. To the contrary, this was just the beginning. The genie had effectively ditched the bottle; she flies freely among us still.

Images of a technologically advanced, utopian society were not, of course, produced solely by Republic Pictures in the 1950s. A young commercial illustrator from Detroit named Arthur Radebaugh did probably as much as anyone to foment the collective notion that soon enough robot maids, flying cars, and, yes, rocket pack–powered mail-men would be touching down in a neighborhood near you.

Radebaugh got his start doing commercial work for the automotive industry. Some of his earliest illustrations already suggest the coming futuristic cityscape. A 1939 drawing throws a boxy Dodge Luxury Liner in front of a twisting monorail far more grandly intriguing than anything Disney's Imagineering crew ever dreamed up. In a later piece, a car that's part Buick, part space pod zips away from a hovering glass bubble of a bachelor pad. Like much of Radebaugh's work, the image takes ornate deco flourishes and transports them to an imaginary, not-so-distant future. Jared Rosenbaum, who helps run the Palace of Culture Web site that features an exhibit of Radebaugh's work, is writing a book about the illustrator, and calls Radebaugh an "eccentric avatar of the future."

In 1958 Radebaugh launched a syndicated newspaper cartoon titled *Closer Than We Think*, which the illustrator described as "halfway between science fiction and designs for modern living." The strip tackled subjects such as "Mining on the Moon"—which evidently requires an enormous green spidery-looking contraption—and "Electronic Home Library," depicting what could be taken for an upper-middle-class home today, only Dad's stylishly clad in a narrowly cut suit and there are shelves packed with ornate hardcover books. In "Rocket Mailmen," the boyish, clean-cut letter carrier arrives at the door of an apron-wearing young wife after activating the single-tank propulsion system on his back. A small dog yaps behind him. The jet exhaust trails to the neighbor's yard, suggesting the mailmen uses his device mainly for quick house-to-house hops. This was the future—in all its fluorescent paint and airbrushed glory—and it sure beat schlepping that mailbag around town. At its peak, Radebaugh's *Closer Than We Think* column reached about nineteen million newspaper readers.

Though it's impossible to know if any of those readers controlled the purse strings that financed new military developments, it was around the time of "Rocket Mailmen" that the U.S. Army began to reconsider whether helicopters were the best way to transport troops. The same year that Radebaugh's strip debuted, the Reaction Motors Division of New Jersey's Thiokol Chemical Corporation won a government contract to look into what was then being called a "man-lifting device."

Under code name Project Grasshopper, Thiokol engineers devised a system whereby a soldier could ignite a solid propellant cartridge, which, ideally, would lift him high and far enough to, say, clear a creek or scale a hill. They called it, modestly, the Jump Belt. A company press release from June 1958 announced: "The infantry soldier of the future … will be able to achieve an element of mobility he has never enjoyed before. The jump belt will enable assault troops to cover vulnerable areas in minimum time. Some specific uses which have been described include leaping from landing boats and dash for cover, attack pillboxes with flame-thrower by leaping over them and pressing home the attack from behind, and in general cover difficult terrain with comparative ease."

Creative grammar usage aside, the army was not impressed, offering a subsequent contract to explore stability issues not to Thiokol but to a California company called Aerojet-General, which had previously built rocket motors for World War II planes. Aerojet would eventually go on to be a big player in the space race of the 1960s, doing engine work for the Apollo moon mission.

Aerojet's AeroPAK was powered in much the same way that Thomas Moore had hoped would get his Jet Vest off the ground. In fact, U-shaped fuel tanks seem to be the major design difference between the two devices; otherwise, they both used hydrogen peroxide as rocket fuel and a silver catalyst to transform the substance into superheated steam in order to generate the necessary thrust.

Although it is generally believed among technohistorians that Aero-jet successfully flew its AeroPAK during brief test skips, the results were not great enough to earn funding for additional development. In the end, said Mark Wells, who has studied the work of both companies, "neither Thiokol or Aerojet was able to demonstrate suitable control." Yet again, the jetpack was tossed onto the scrap heap of history, somewhere between a teleportation chamber and a universal translator. And there it would've likely remained were it not for one supremely obsessed, preternaturally driven, bow-tie-wearing midwesterner.

Wendell Franklin Moore was born in Canton, Ohio, on March 6, 1918. His mother owned a small dry-cleaning business; his father worked in a steel mill. When Wendell was a young boy, his dad's persistent gambling and drinking led to his parents separating. Soon after, his mother moved Wendell and his older sister into their grandmother's house. "He never really spoke about his dad after that," Moore's daughter Carolyn Baumet told me. "I never met my grandfather."

From an early age, Wendell seemed keen on escaping the dead-end factory life he saw around him through blue-collar aunts and uncles who took jobs at local plants just to barely scrape by. At the remarkably young age of ten he started working at the airport, washing and painting the propeller planes and even performing minor repairs. "He grew up loving airplanes from when he was a baby," Baumet says. "He lived, ate, and slept airplanes."

That passion eventually led Wendell to enroll in engineering courses at Tri-State College in Indianapolis, but he soon grew restless and dropped out before earning a degree. Instead, he went to work in the propulsion department of a local aerodynamics firm. At that point in his life, Wendell Moore was not much different from any other nineteen-year-old male: fascinated by the nascent flight industry and scanning the newspaper for his daily dose of the rocket fuel–soaked adventures of Captain Buck Rogers.

One night he drove out to a social mixer in nearby Hillsdale. He wasn't a particularly smooth operator as far as women were concerned, but that evening he found himself chatting up a small-framed farm girl with short red-brown hair.

By 1940 Norma Alice Watkins and Wendell Moore were married and settled into a quiet middle-class existence in a little town called Bryan, Ohio. Before too long a baby girl was born, but at eighteen months she died of complications related to chickenpox. Perhaps it was the trauma of this untimely death, or maybe it was just Wendell's insatiable wanderlust, but the couple soon moved back to Hillsdale, where another daughter, Carolyn, was born in 1942.

The simple life wouldn't last. Wendell, now with a few years of aeronautic work under his belt, was once again ready for a challenge. The air industry was beginning to boom, and Wendell hated the idea of fiddling with formulas in the sticks when he could be part of the future of flight, working with rocket-powered planes.

And so, in 1945, the family packed up yet again, this time heading east to the Niagara Falls plant of the Bell Aircraft Corporation, where Wendell had taken a job as a propulsion engineer. The company had already gained industry-wide acclaim by producing the first commercial helicopter. A year after Wendell was hired, Bell began work for the U.S. Air Force, developing the supersonic airplane technology that would eventually lead to their speed of sound–busting, needle-nosed, pumpkin-orange X-1 and, later, the X-2. Now this was Moore's idea of fun—he joined the X-1 team, splitting time between Niagara Falls and Edwards Air Force Base in California's Mojave Desert, where the planes were tested.

The restless, would-be renegade partial to horn-rimmed glasses, a flattop haircut, and bow ties had finally found a lifestyle that agreed with him. Never in one place for long, Moore dove headlong into his work, thriving in those heady days of engineering when any and every innovation of the air seemed just on the horizon. "He was a gregarious person," Bob Rohrer, an instrument technician who worked with Moore on the X-plane projects told me. "You couldn't help but like him." "He was a very likable guy," seconds another of Moore's Bell colleagues, Bob Roach.

Not everyone was convinced. Moore's daughter, for one, was not happy with the new job. "My whole school life we moved back and forth a lot—that was a sore point for me. Making friends and being part of a high school was very difficult for me."

Father and daughter did find one way to bond, however, by flying Moore's Stinson Voyager prop plane between Niagara Falls and Canton to see Carolyn's grandmother. "I was his copilot," Carolyn said. "That was our way of communicating—through our love of airplanes and flying. He taught me to navigate and to fly. He let me hold the wheel."

Those trips were a rare treat. More often, Wendell was so engrossed in his work that he and his boss, Bill Smith, whom Moore called "Smedley," were known to stick their heads within inches of the X-1's rocket engines during test firing to get a better look. "They were looking to see if there was any leakage," said Roach. "That was pretty dangerous."

In those early days of the space race, there was tremendous pressure to perform, and the Bell guys needed an after-work outlet for letting off steam. An annual Christmas Eve party hosted by the company's technicians included a robust raw bar and smuggled booze. When attendees were suitably liquored up, out came the scissors: it was time for the traditional snipping of all engineer neckties. "The techs would get drunk and fire off rockets," Bob Roach laughed. It didn't take a holiday to prompt a party; even minor successes called for a round of drinks at the local watering hole on Niagara Falls Boulevard.

Moore thrived in this work-hard, play-hard corporate culture. He smoked two packs of Winstons a day until his asthma became so bad he was forced to quit. He drove a flashy red-and-white Ford convertible. Every day at lunch he drank a beer, often dropping in a raw egg for health purposes. "He loved his beer, it's true," Carolyn Baumet says, sighing. "He was quite a beer drinker."

Moore's fondness for good times did nothing to dim his limitless gusto for work. And things were going extremely well for the X-plane team. In 1947 Bell hired a dashing young pilot named Chuck Yeager to test out the merchandise. The results were nothing short of spectacular; his first year on the job, while flying Bell's X-1 plane, Yeager smashed the sound barrier, something no pilot had done before.

Lodged beneath the belly of the sprawling, ninety-nine-foot-long B-29 Superfortress like some sort of winged embryo as the larger plane drifted along, when the X-1 was birthed into the air, Yeager would let the craft glide for a moment or two before firing the four rockets, at which point all hell would break loose. In a good way. In 1953, just a few years after first flying the X-1, Yeager rewrote all existing flight records when he cruised at more than double the speed of sound, or a cool 1,650 mph.

But while piloting a test flight later that year, Yeager's blindingly bright career nearly came to an abrupt end. On the brisk morning of December 12, 1953, just after dawn, Yeager climbed into the cockpit of the next generation of supersonic planes, the X-1A, intent on breaking the new record of twice the speed of sound. He took the silver bullet up to 40,000 feet, then 50,000, 60,000, and finally just over 76,000. His instruments registered his velocity increasing from Mach 2.3 to 2.4 and even 2.5.

And then—blackness. Suddenly, the pilot's controls were useless in his hands. Nothing he tried had any effect on the X-1A as it tumbled, first flipping sideways, wing over wing, and then in an inverted spin, tail over nose. Yeager lost consciousness as the plane's altitude dropped dramatically, its shadow expanding across the sand of the Mojave Desert. Then, at about 25,000 feet, Yeager awakened and woozily fought for control of his vessel. Finally, the X-1A's instruments came back online, and the captain steered the plane toward a smooth landing. "I, uh, got in bad trouble," Yeager told ground control, as he glided toward earth. On recordings of his communications his voice is cracked and halting. "I can't say much more—I gotta save myself. I don't know whether I tore the thing up or not. Christ! Boy, I'm not gonna do that anymore."

Not lost in all the excitement was the fact that Bell's creation had in fact set a new air-speed record. The crew retired to a nearby bar to celebrate. Anyone wearing a silver bell attached to their lapel was treated to free drinks all night. Standing in the cigarette haze, the clinking of glasses providing a cheery soundtrack, Wendell Moore was already looking into the future.

It's funny where certain ideas come from. Even before Yeager's wild ride, Moore had been thinking about jetpacks. He came of age in the Buck Rogers era and, in the odd moment when he wasn't consumed with Bell projects, let himself daydream about building such a machine. Once, idling in the desert between X-2 tests one day, Moore had taken up a stick and, in front of an audience of a small clutch of colleagues had drawn in the sand a sketch of a jetpack prototype. It looked very much like the device worn by the hero of his favorite boyhood comic strip. All he would need, he supposed, were rocket engines small enough to be worn on a man's back and powerful enough to lift him off the ground.

Yeager's near disaster had given Moore an idea—if the plane's controls faltered above 20,000 feet or so, an alternate source of power would be needed in order to perform at higher altitudes and speeds. Jet engines relied on the atmosphere to generate thrust, but rockets were self-sufficient. What if small rockets could be affixed to the wings, tail fin, and nose, then fired as needed to direct the plane's movement?

After the speed-record party, Moore headed back east to work on his new idea, back to the four-bedroom house on Lake Ontario he'd bought not long after going to work for Bell. It had previously been

owned by a wealthy Italian family and used as a summer cottage, and
it cost Moore eighteen thousand dollars. The place sat on two acres,
some of which Wendell converted into what he called a "skid strip"—
a long wedge of dirt where he could land the light airplane he was
building in his garage. In the summer he'd have his closest Bell cohorts
out for languid barbecues in the backyard near the water. His right-
hand man, the burly, growling Ernie "the Bear" Kreutinger, a Bell technician and former air force pilot, would be there. So would fellow
engineer Ed Ganczak, he of a narrow, steely gaze. Smedley might have
turned up once or twice.

Moore would get up early to light the coals in the barbecue pit
and then hold court all afternoon, a cold beer in one hand, the squirt
bottle of sauce in the other. "I think that's what he did to relax," his
daughter Carolyn Baumet told me. "Just hang out there by the pit,
spraying chicken."

But Moore didn't relax often. There were those minirockets to tend
to, for starters. Working quickly with a small team, the engineer traveled down a path of discovery that the unrelated Thomas Moore was
also traversing in Alabama. The result of his labor was a remarkably
strong rocket about the size of a kitten. At only six inches long and
two inches in diameter and weighing less than a pound, the machinery, a key component of what Bell was calling "reaction controls," was
capable of providing up to 300 pounds of thrust. "To drive the rocket,
he used hydrogen peroxide," Bob Rohrer told me. "Which is an unstable, you might even say explosive, fluid. You had to be very careful
with it—anything could be a catalyst, even a piece of dirt." The cat-
alyzing force of choice, however, was a motor packed with granules of
potassium permanganate, a chemical substance that can be used to
make fireworks.

The rocket was a resounding success, an engineering masterpiece
even. A piece of technology so reliable, several years later the same reaction controls would be utilized in NASA's Mercury space expedi-

tions, lunar-landing equipment, and, in later variations, for shuttle missions. "The rockets were used anywhere above the atmosphere to position and control the vehicle," former Bell rocket engineer Bill Fleming explained. "In orbit, the only way to orient the thing was with reaction controls."

The *Buffalo News* called the tiny rockets Moore's "most enduring contribution to space technology." Bob Rohrer went the *News* one better: "I think Wendell Moore's best invention was the reaction controls."

Jetpack lovers might disagree. Emboldened by his new standing at Bell and with visions of Buck Rogers still dancing in his head, the nearly forty-year-old Moore believed he was now ready—finally, finally—to turn fantasy into reality.

Though the more conservative Smedley was less than thrilled, Moore forged ahead with plans for a working jetpack. It was, in fact, partly his superior's dismissiveness of the project that drove him. "Wendell always wanted to fly up to Smedley's second-floor office and thumb his nose at him," Bob Roach said.

With no budget to speak of, that seemed like a remote possibility. Nonetheless, in the shadowy corners of Bell's mile-long warren of expansive airplane hangars and cluttered offices, Moore and Company got to work. Stealing equipment and tools from other departments, they began to assemble what only a few years earlier had been a doodle in the sand. "He didn't get much support from management, but he kept plugging," said former Bell engineer Charles "Ed" Satterlee, who worked for the company for thirty years. "I had a desk right next to him. He would ask me questions all the time, but I was always too busy. Moore had his own little group—they had meetings every morning."

If there were setbacks, and there were setbacks, Moore simply brushed them aside. "I found Wendell Moore to be an indomitable spirit," a former vice president at Bell, Hugh Neeson, told me. "He was dogged. That generation of engineers, when faced with a problem, would just go and fix it with skill and intelligence."

That and the odd lunch-hour bullshit session at the local firehouse-cum–beer hall. "C'mon, kid, let's go have a couple of beers" was how it invariably began, Moore's enthusiasm impossible to resist. Soon there'd be an egg floating in suds like a soggy white Cyclops eye, while Moore loaded his beef on weck with horseradish in astonishing amounts. Apparently, the horseradish helped with his asthma—he called it "Jewish Dristan."

That unlikely mixture of skill, intelligence, and horseradish worked wonders: within a few months, the team had hammered out a crude prototype. Machined and manufactured almost entirely out of stainless-steel parts left over from the burgeoning Bell aerospace department (as well as handlebars from an old Harley, random parts from someone's kid's bicycle, and duct tape), it looked like hiking gear yanked out of a distant future.

Three Tylenol capsule–shaped tanks ran practically the length of the pilot's back. The middle container held the liquid nitrogen that would be used to pressurize the hydrogen peroxide in the flanking tanks. The peroxide was distilled to a concentration of 90 percent, or about 87 percent stronger than the stuff you used to bleach your hair during a drunken college dorm party.

The tanks were welded to what the Bell team referred to as a corset made of fiberglass from a mold poured over Moore's back and fitted to his lean, five-foot-nine frame. Jutting forward from the corset were two steel arms at the ends of which was a pair of throttles similar to those you might find on a motorcycle but perpendicular to the ground, rather than parallel. The right throttle controlled the gas flow; the other manipulated the O-rings—or what are now called, incredibly coolly, jetovators—fastened to the tip of the nozzles. With a twist of the left hand, these jetovators could be angled forward or back, thus sending the real-world Buck Rogers up, up, and, briefly, away.

The lucky soul operating this machine would be strapped into the corset by a single automobile seat belt around the waist, a strap across the chest, and two more around the upper thigh. For this reason, Moore called the contraption a Rocket Belt.

One frigid day in late 1960, Wendell Moore strode into a massive hangar on Bell's Niagara campus. He wore a shiny firefighter's flame-retardant suit, the hems duct-taped down into a severe taper for added safety. He had on dark leather work boots, a racing crash helmet, and, as always, his horn-rimmed eyeglasses. He was ready for liftoff.

Early kinks had been worked out on a test stand that would allow the pilot to go up and down (but not blast off to, say, Cleveland by mistake). One engineer had had the arms of his suit coat blown to tatters by the superheated steam of the jet exhaust. This led Moore to reposition the nozzles farther away from the pilot's body, which, in addition to staving off severe skin burns, increased stability.

Funding had been secured, too. When Aerojet-General's contract expired in 1957, the U.S. Army's acronym-happy Transportation, Research, and Engineering Command (TRECOM) circulated a call for demonstrations of a Small Rocket Lift Device (SRLD). The timing could not have been better for Moore and his gang; they nabbed the gig. The total winnings were twenty-five thousand dollars, which helps explain the DIY approach taken by Bell.

Moore hooked a tough nylon safety rope to a harness he'd attached to the Rocket Belt, behind the pilot's head, for preliminary flights. He bounced slightly on the toes of his boots. His team of ten men surrounded him: two photographers, Bear Kreutinger, Ed Ganczak, assorted engineers, technicians, and the company doctor. Wendell Moore flashed thumbs-up. Then he twisted open the throttle valve.

One hundred and thirty decibels worth of superheated steam—louder than a Motorhead concert—shrieked through the jetovators. The duct-taped trousers registered a hyperspeed flutter as the hot air hit them at a rate of 3,000 feet per second. Moore gripped the controls until his knuckles turned blue. And then he was dangling above the earth. Without a propeller, parachute, or wings. Just a man and his machine. Flying.

Well, hovering, really, but who's about to split those hairs? The flight lasted less than ten seconds but yielded an incomparable elation that would last a lifetime. Years later Moore would tell the hosts of a radio program called *Off the Wing,* "Oh, boy, ever since I was a kid I wanted to fly—and the Rocket Belt has real possibilities." He must have sensed those possibilities that first winter day in the hangar, and with them an inkling that his boyhood dreams might yet come true.

Subsequent indoor tests brought mixed results. It seemed not quite all the kinks had been fixed. For every smooth ride, there was a wall-denting misadventure. There remained issues of stability and valve control. Some Bell colleagues began to wonder if the Rocket Belt, though obviously an exciting futuristic gadget, was such a good idea after all. "He whacked himself against the walls a couple of times, and everyone was saying, 'What the hell is he trying to do?'" former Bell engineer Bill Fleming said.

Luckily, Moore was tenacious and simply refused to be beaten. "Wendell Moore pursued development of the Rocket Belt until it became a practical vehicle," the onetime head of Bell's Aerodynamics and Propulsion Department, George Melrose, told me. "In the environment of a large corporation this involves many things: selling management on the idea, looking for customers, and putting together a group of competent and dedicated people in both the analytical area and experimental. Moore was the captain of the team."

The captain, while doggedly pursuing his dream, somehow also kept his sense of humor. A deeply devout Catholic who, for a time, wanted his daughter Carolyn to be a nun, Moore was once asked by a colleague what the pilot should do if the Rocket Belt engine fails midflight. He waited a beat and then responded: "Grab a rosary."

Unfortunately, there was no time for that during a tethered test one morning later in the winter. Hovering about ten feet over the hangar's concrete floor, Moore shut off the fuel flow, expecting to be lowered to the ground. But no one in his crew had noticed that the nylon safety rope, scraping against one of the engine's razor-sharp brackets, had been snipped during the operation. Moore crashed to the pavement, shattering his left kneecap. He would never fly the Rocket Belt again; the company's insurance policy wouldn't allow it.

Hobbled and denied the chance to experience a truly free flight, Moore spent the rest of the winter stewing, tormented both by the grounding and by a leg-long cast that stopped just short of his groin. "That thing itched so bad," Carolyn, his daughter, said. "I can still see him sitting there, swearing up a storm and hollering at my mom to get a coat hanger—he straightened it and used it to scratch." When that didn't work, he'd drink pickle juice or anything sour to cause a relief-inducing shiver. "He was as mean and cross as two sticks," Carolyn continued. "He was not meant to be doing nothing."

Compounding the misery was the idea that he'd come so far with the Rocket Belt and had been so close to breaking free of the hangar's constraints. As Carolyn noted, "It broke his heart when he found out that he couldn't fly it anymore, because that was his baby. It really did a number on him." Bell colleague Bob Roach told me: "I think that was his greatest disappointment, that he never got to make a free flight."

Tensions increased at home, where Wendell and Norma were already struggling with the birth of an epileptic son a few years earlier. Carolyn recalled many nights when the kitchen light would remain on deep into the night as her parents, sometimes loudly, discussed the kids, work, the future.

Still, fortune continued to smile on the Rocket Belt project. Two days before Moore's accident, a young engineer had been rehired by Bell to act as backup pilot.

Harold Graham was an energetic twenty-six year old with the build of the ice hockey player he was. He'd been hired by Bell a couple of years earlier, taking on the undesirable graveyard shift in the rocket-testing department of the Mercury project. For a year and a half he punched the clock and worked for twelve hours a day, seven days a week. Quickly burned out, he soon resigned and was casting about for something better when the phone rang one day. It was Bell Human Resources—there was a new opening. And it was a day job. "So I went for it," Graham said in a telephone interview. "Turns out it was a lucky break for me."

And how. Graham picked up where Moore had left off, with a few minor differences: when the company doctor checked him before his first tethered hangar flight, Graham's blood pressure spiked at 140. But he soon got over his fears. Thirty-six tests later, the Rocket Belt team was ready to move outside.

On April 20, 1961, a date that for jetpack obsessives carries as much significance as July 20, 1969, does for space nuts, and fifteen days before Alan Shepard first climbed aboard the Mercury, the Bell crew gathered early in the morning on a stretch of turf at the edge of the Niagara Airport. Graham maneuvered into his 140-pound pack. He exhaled deeply, his breath billowing in the cold air. Cars jammed the nearby roadway, drivers straining to catch a glimpse of this comic-book page come to life.

Then Graham was suddenly in the air, toes flexed a foot and a half above the grass, which flattened under the exhaust's force. The pilot nudged the jetovators. He flew forward. Three, four, maybe five miles per hour. The photographers ambled along in wool suits, clicking madly, keeping pace. In Bell's film footage, Graham is initially obscured by the clouds of white steam generated by the hydrogen peroxide meeting the low temperature of the air. But as he floated now, the smoke cleared to reveal a determined, if not completely relaxed, pilot kicking gravity's ass like it had never been kicked before. One hundred and twelve feet later—or eight less than the Wright brothers' maiden voyage—Harold Graham's boots reconnected with Earth. There would be many, many drinks that night on Niagara Falls Boulevard.

The team sobered up quickly. Only months remained on the army's contract, and much work was to be done if Bell hoped to get an extension of two hundred thousand dollars. Which, of course, Wendell Moore badly wanted, if only to have time and resources to improve on the Rocket Belt's twenty-one-second flight time, a cruel constraint of physics due primarily to the amount of fuel a pilot could comfortably hold on his back. No question, what Moore had come up with was impressive—a man could now come as close as he'd ever been to flying like a bird—but what was the average soldier going to be able to accomplish with a mere twenty-one seconds in the sky?

As it was, the nitrogen tank held two pounds of fuel, whereas forty-seven pounds of hydrogen peroxide was stored in the other compartments. So the Rocket Belt captain was already tinkering with possible alternate fuel sources and thinking about ways to lighten the pack to increase air time.

Meanwhile, Graham could now launch himself up a thirty-foot hill, leap over a twelve-foot stream, and slalom between ski flags like a cross between Swiss ski legend Silvan Zurbriggen and a Hadada Ibis. He was, in other words, ready for his close-up.

The first public demonstration of the Rocket Belt occurred on June 8, 1961, at Fort Eustis in Virginia. Graham and Bell's inner circle—Moore, Kreutinger, Ganczak, and Dr. F. Tyler Kelly—traveled down. The army's point man, a civilian named Robert Graham (no relation to Harold), met the crew and marched them over to a patch of turf that was empty, save for a single parked army truck. Ringing the field were several hundred high-ranking officers and their guests. The plan was for Graham to lift off, elevate above the truck, fly over it, and land. Without breaking anything.

By this point Moore knew he had a reliable machine and a talented pilot. His nerves were not so jittery that a pack of Winstons couldn't do the trick. He hung back in the crowd and waited for that now familiar shriek.

And then it came. Graham drifted high above the vehicle, the black dot of his helmet providing the point on an inverted exclamation mark of a man. When he floated softly back to Earth, the crowd met him with booming applause. Buck Rogers was finally among them. Graham grinned and threw an unrehearsed salute. "By the time we got back to the hotel," Bob Roach said, his breath still catching all these years later, "the phone was ringing off the hook," with calls from local newspapers and the Associated Press.

"Knapsack-Like Jet Enables Man to Fly," the *Hartford Times* breathlessly but wrongly asserted. "Flying Belt Rockets into Reality," exclaimed the *Waukegan News-Sun*. Many publications pushed the Buck Rogers connection: "Man Flies Like Buck Rogers Now," "Buck Rogers' Space Belt Becomes Reality," and, most poignantly, "Buck Rogers Era Here." Others turned to more organic analogies. "Old Dream of Flying with Birds Now True," exclaimed the *Philadelphia Bulletin;* "Belt on Back Bolts Birdman," the *St. Petersburg Independent* alliterated.

Invoking the flight of birds in this way effectively aligned Graham with myths like Icarus and men like da Vinci and Alberto Santos-Dumont. Santos-Dumont was the eccentric, well-tailored Brazilian who, in the early twentieth century, flew lighter-than-air balloons around the Eiffel Tower and was known to park his vessel on the roof of the restaurant at which he was lunching. Once, reminiscing about his childhood, he said, "I would lie in the shade of the verandah and gaze into the fair sky of Brazil. Where the birds fly so high and soar with such ease on their great outstretched wings, where the clouds mount so gaily in the pure light of day, and you have only to raise your eyes to fall in love with space and freedom."

And just like Alberto Santos-Dumont, Hal Graham was suddenly a celebrity; the Belt, as it was known, was a hit. The *Toronto Star Weekly* arranged a photo shoot, wherein Graham played your average commuting husband, and Carolyn stood in as the doting housewife. The paper shot scenes at the Moores' place of Graham leaving for work in the morning, kissing his wife good-bye, walking out onto his back porch, and strapping on his jetpack, as if it were tomorrow's Packard.

Soon, the biggest names in the government were clamoring for a look. At a Pentagon demonstration Graham flew once in the morning and again in the afternoon in front of an estimated six thousand officers and assorted VIPs. For that demo, the young pilot leaped a parked military sedan, traveling about the length of a football field before touching down. The next day's *New York Times* carried a report of Graham's thirty-foot-high trek on its front page. A few months later, Moore and his crew were called to Fort Bragg for their biggest moment yet, a ship-to-shore operation for none other than President Kennedy. In grainy clips of that historic hop, Graham jumps from a small vessel, blitzing low across the water and whitecaps of McKeller's Lake. And then he's ashore, a spaceman on the beach, saluting JFK, who returns the gesture. *Life* magazine captured the exchange from the president's point of view, the familiar silhouette looking upon American ingenuity at its most creative. The *Buffalo Evening News* wrote that "Mr. Kennedy was described by an Army Officer sitting near him as 'wide eyed and open mouthed—just like a kid.'"

"My favorite memory has got to be the JFK flight," Graham told a PBS affiliate years later. "I mean, the president of the United States— where do you go from there?"

One answer: to a demo in Phoenix for then Secretary of Defense Robert McNamara. Another: nowhere fast. Well, at 40 mph, anyway. Although the public and certain corners of the military were understandably smitten with Moore's invention, there was one serious design flaw yet to be worked out—that frustratingly short twenty-one-second flight time.

Bell had been awarded a contract to demonstrate the concept of its flying machine, but until the company could improve on the Belt's flight duration, the government wasn't about to cough up any more dough.

Wendell went back to his Niagara Falls drawing board, his fifteen minutes of fame behind him. Harold Graham went on to careers in teaching, as an accountant, town justice, engineer, computer programmer, and, currently, charter-plane pilot. Now seventy-three years old, he lives with his girlfriend fifty miles west of Knoxville, Tennessee, in a town so small you can mail him a letter without including a street address and he'll get it.

"The good part is that I did it for the first year and a half we went public—it's nice being number one," Graham told me. "If the Wright brothers were the first to fly, who was the fifth to fly?" It's a point well taken, but the truth is that in this case, it was not the first Bell Rocket Belt pilot that history has remembered best of all. While Moore was tirelessly investigating ways to keep his baby airborne, Bell's PR Department was busy fielding calls for demonstrations. By the end of 1962 the army's interest might have been cooling, but requests to book a flying man were flooding the home offices in upstate New York. For a while it seemed every stadium, fairground, flight expo, and mall opening wanted the Rocket Belt. Demand was so high that Bell hired and trained three new pilots, Gordon Yeager, Peter Kedzierski, and Robert Courter Jr., and sent its teams off with freshly made machines to satisfy customers all over the world.

In 1964 alone, Bell flew at exhibitions in Honolulu; Copenhagen; London; Ontario; Sydney, Australia; and for 170 quickie demonstrations at the New York World's Fair. During 396 total flights, an estimated two million people were exposed to the wonder of a man defying gravity. At the International Aviation and Space Exposition in Paris, Kedzierski and Courter flew for more than one million slack-jawed spectators over a nine-day burst of future-looking fantasies. By now pilots were able to crank the Belt up to 60 mph and scale a fifty-foot tree, which helps explain why *Paris Match* magazine dubbed the dynamic duo "the bird men."

That same year, 1964, Moore made the most prescient hire of his life. Bill Suitor was the youngest son of Moore's Youngstown neighbors. As soon as Bill was old enough to start a lawn mower, Moore had him over to take care of the yard and perform other odd jobs around the house.

Born into a blue-collar family—his dad worked for the gas company his entire life, rising from meter reader to manager over a forty-year career—Suitor was a smart but somewhat aimless young man. After graduating from high school, he enrolled in a local architecture program but quickly realized his heart wasn't in it. So he did what any guy his age with an adventuring spirit might've done: he paid a visit to the army's recruitment office. "There was a waiting period before you could join," Suitor told me. "I did everything but take the oath. But my father wasn't happy that I was dropping out of school to join the army."

To placate his pops and perhaps also as a last-chance alternative to enlisting, Suitor started hitting up Moore for a job at Bell. The youngster bumped into the inventor in town, at the grocery store, all over Youngstown, always offering the same refrain: "When are you going to hire me?" Some days he'd wander over to his neighbor's garage where Moore was building his airplane. "I remember going in there and watching him. When you're a kid, you know, it's really fascinating. And then we hear he's working on a jetpack or a rocket belt or whatever, so everybody just thought, 'This guy's nuts.'"

Moore was certainly locally anomalous as a renegade scientist with a wicked sense of fun in a town of modest, straitlaced factory workers. On New Year's Eve, the engineer enjoyed honoring his German heritage by visiting neighbors with a bottle of whiskey, a shot glass, and a bowl of sauerkraut, asking them to toast to health and prosperity. "He was the talk of Youngstown," his daughter Carolyn said. "He'd always come home schnockled," thus prompting a fresh round of late-night kitchen conferences with Norma.

One day while waiting on word from the army, Suitor was unloading a trailer full of mufflers and tailpipes at an auto parts shop where he occasionally worked. His mother called and told him, "On your way home, stop by Wendy's," using Moore's nickname. "He wants to talk to you."

Wendy was in his office, smoking by the window and drinking a beer. It was late afternoon on a Friday, the sun streaking into the room. Moore didn't mince words: if he passed a physical on Monday, Suitor would be hired as the newest Rocket Belt pilot. The U.S. Army was out of luck—it was losing its newest recruit.

By the following Monday afternoon, in one of Bell's cavernous hangars, Bill Suitor was getting acquainted with a bit of machinery he'd eventually know better than anyone. Bell had brought him in to establish the idea that the belt could be mastered by any kid off the street, regardless of previous piloting experience. Suitor had never even been a passenger on an airplane. "Word was sent down that I passed my physical," he told me. "And I signed this, that, and the other thing. I was nineteen years old. I didn't know what the hell I was signing." His salary was $147.50 per week.

It wouldn't be hard to argue that the kid was underpaid. Following a series of about sixty tethered and then free test flights, he was deemed ready for his first public event. On a scorching August day, wearing the same wool suit he'd worn to high school graduation and lugging a pair of suitcases stuffed for a two-month adventure, Suitor climbed aboard a Lockheed Electra turboprop plane bound for Chicago. Leaning on the kindness of a few strangers, he fumbled through O'Hare and made his connection west.

A few days later, Suitor was to debut during a nighttime show at the California State Fair in Sacramento. The plan was to blast off from inside the horse-racing track's oval, whoosh by the dumbstruck spectators in the grandstand, and land stage left next to Bob Hope, near the orchestra pit, facing the crowd. "I'm as nervous as a whore in church," Suitor recalled. "Flying the Rocket Belt is an experience in self-preservation, first of all—flying it in the dark is kind of like falling down a mine shaft."

The nineteen year old took off and immediately knew there would be issues. For one thing, he could barely make out the stage and was outracing the spotlight operator. This turned out to be a stroke of luck, however, when the bright bulb was instead aimed at the spot on stage where Suitor was to land. Now he had it in his sights.

But then the buzzer inside his helmet started blaring—the timer counting down ten seconds before the fuel ran out. Wrestling with his sky-bound bull of a machine, Suitor came in too low on his descent, the belt's exhaust blowing hard over the crowd and orchestra. Sheet music whipped into the air like wedding confetti, the tuxedoed players scrambled for their very lives. "It was panic, pandemonium," Suitor said. Somehow, he managed to wrangle the thing up and over the lip of the stage, landing amid a hurricane of music stands, fluttering pages, tangled microphone chords, and a befuddled Bob Hope. He checked—yep, he was still in one piece. Bell's unblemished safety record remained intact.

Safely back in Niagara Falls, Suitor fell happily into life in the hangars. The scrappy do-it-yourself aesthetic of Bell's Rocket Belt team appealed to his young, spirited, anything-goes attitude. When Wendell Moore, still searching for ways to improve his design and woo investors, affixed rockets to a plastic office chair, Suitor was first in line to fly it. The two-man flying platform and something called a "lunar pogo stick"—which looked roughly like what it sounds like—were other works in progress that fearless Bill fired up.

When not testing these flying found-art objects or jetting off for a public demonstration, Suitor idled in the marketing department or fetched parts, bicycling from hangar to hangar, amid endless shelves piled high with assorted stainless-steel contraptions to be machined into magic.

Ironically, whereas in the air Suitor was quickly becoming the undisputed champion of the belt, on the ground he was a bit of a klutz, so much so, in fact, that he developed a nickname appropriate to a chronic bumbler: Peter Sellers. "You know, it was comical after a while," he told me.

"Once I made this part for a yaw handle, and I had to take these parts I'd made down to the weld shop and have them welded and heat-treated. So I put them in the basket on the bicycle, and it was a beautiful summer day and I'm riding back on one of the roads and I cross the railroad tracks that came through the plant, and when I did one part fell out of the basket through the wire mesh. The bicycle tire ran right over it, and as I went over the track, it bent and formed itself around the tire. And it was going click-click-click-click until I stopped. I went, 'Oh, shit!' But I was able to bend it back."

Suitor was obviously doing plenty right, too. When Hollywood inevitably came calling, he was dispatched to help none other than James Bond look cool. In what remains one of the most profound pop culture touchstones for jetpack junkies the world over, Sean Connery's Bond used Bell's Rocket Belt to befuddle the bad guys during one of his signature heart-racing pretitle sequences, this one for 1965's *Thunderball*. Except it wasn't, of course, Connery at the controls, hopping off the ledge of a wilderness chateau in the Bahamas, the better to elude SPECTRE's goons. It was the intrepid Bill Suitor, by then a twenty-year-old veteran of hundreds of twenty-one-second blastoffs, gingerly touching down just beyond the fortress's walls, where Bond's girlfriend du jour was waiting with the requisite getaway Aston Martin. At which point Connery takes over, removes his helmet, and folds his flying gear into the car's trunk—or boot, if you prefer. "No well-dressed man should be without one," he calmly explains to his dark-haired temptress.

Years later, no matter their sartorial state, Bondophiles couldn't agree more. In two separate 2006 polls—taken by the *New York Post* and *Popular Science* magazine—ranking the overall badass-ness of the spy's legendary gadgetry, the jetpack came in first and second. It beat out such classic Bond accessories as a robot dog, a killer briefcase, a laser watch, and a camera gun.

Bill Suitor's on-screen career continued the following year when producers of the hit TV series *Lost in Space* summoned him and Bell colleague Gordon Yeager to the rocky, rustic Death Valley set. Taken together, Buck Rogers comics, *Thunderball*, and *Lost in Space* represent three of the five pillars of pop culture history that have effectively secured the jetpack's place in the public's consciousness. The other two pop projects would not be made for many years yet, but if Bond has proved anything, it's that his back catalog has serious legs. Sean Connery had elevated the jetpack's profile considerably, holding it aloft for ensuing generations to marvel at—and to covet. *Thunderball*, a jetpack splashed across its promotional posters, was the highest-grossing Bond film in the series until the eighth installment, 1973's *Live and Let Die*, surpassed it. If total box office dollars are adjusted for inflation, *Thunderball* remains the most lucrative 007 adventure to date, with nearly $925 million under its Rocket Belt, almost $100 million more than the next closest film. It doesn't take too much of an imaginative leap to think that his scene-stealing, skyscraping costar is one spectacular reason.

While his staff rubbed shoulders with the glamorous and the beautiful in the land of make-believe, Wendell Moore pursued his goal of a real-world jetpack that the average commuter could, like Harold Graham in the *Buffalo News* spread, take to work. "That was his dream— that it would become like a second car," his daughter Carolyn told me. "That and to have the army pick it up. It was like a vehicle of the future for him."

The immediate future, unfortunately, was not looking good. Desperate to increase flight time, Moore had begun to phase out the rockets that had served him well, instead turning to jet technology, which could potentially use the atmosphere's airflow to greatly enhance a pilot's journey.

Around the time that Suitor was doing stunt work for James Bond, Moore was holed up in his Niagara Falls office hammering out a proposal on his next-level machine. The Jet Belt plan was sufficiently enticing to the military's Advanced Research Projects Agency that it dropped $3 million in Moore's lap and told him to get to work.

Bell wasted no time in commissioning a prototypical engine. Michigan's Williams Research Corporation had already been dabbling in small but extremely powerful jet engines, building them for a fleet of Canadian aircraft. For $2 million they delivered a 67-pound gem of engineering acumen. The WR-19 gas turbine engine was about the size of an award-winning watermelon and could generate an unprecedented 430 pounds of thrust. *Aviation* magazine was mighty impressed, noting enthusiastically, "The 'Buck Rogers' belt now has a jet engine that gives it longer range for a whole new variety of missions. Who knows, someday it may even become a commuter's vehicle."

There were just a few problems. Ready to fly, the Jet Belt weighed some 170 pounds. The bulky design included two plastic tubs to hold the jet fuel and a thick pair of nozzles that protruded on either side of the operator's head like baby elephant trunks. When cranked on, it throbbed at an ear-crumbling 140 decibels (the New York City Subway hits only about 90 decibels).

Though Bell and Williams jointly successfully demonstrated a four-and-a-half-minute flight at a Virginia army base while working toward an hourlong final product, Bill Suitor and others at Bell could already read the writing on the wall. For one thing, the camaraderie and excitement that surrounded the Rocket Belt's early days were gone from the sprawling campus, replaced by whispers of anglings for a $10,000 pilot bonus and a doomed future. Suitor was largely left out of any Jet Belt planning, as demonstrations were carried out by the dark-eyed Moore crony Bob Courter. "I think what happened with the Jet Belt is, those in the middle of it knew there was an end in sight and they were going to get what they could," Suitor told me. "Looking out for No. 1—which I guess is human nature."

If there is a moment when the dream died for the young pilot it was one day when he was delivering some drawings from the art department to marketing. Walking between two buildings he heard a noise like a giant vacuum cleaner.

"I looked around the corner, and there was the God-damned Jet Belt hovering. It was like somebody kicked me right in the stomach, you know? I was supposed to be a part of this, we were all in this together, and here they were flying the son of a bitch and they didn't have the common decency to invite us to come watch. I could have sat down and cried—it really, really hurt. There were some traffic cones on the grass, and Courter was about six feet in the air. I stood and watched him for a few seconds and then left, in disgust, really. I never saw it after that."

Suitor was not alone—the project quickly fell apart. Disenchanted by the machine's many drawbacks and distracted by the growing conflict in Vietnam, the government soon lost interest.

The end came quickly for Wendell Moore, too. On May 26, 1969, still hoping to drum up funding and enthusiasm, Bell released the first images of Jet Belt flights to the public. Three days later Moore collapsed during a heart attack in his Lewiston home. By the time the ambulance arrived from Youngstown, he was dead. For the ensuing week, upstate newspapers, TV, and radio eulogized their local hero. At the funeral the crowd was so large it poured out of the church's doors, with mourners lining up down the block. "It was crazy," Carolyn Baumet told me. "I don't know where all of those people came from, but they came out of the woodwork."

On the one hand, it could be said that Moore died at the top of his game. The week after he collapsed, he was to have met with none other than Walt Disney to kick around some ideas. On the other, despite his considerable accomplishments with both the Rocket Belt and the Jet Belt, he never could get either machine over the final hurdles. One interpretation of his last days could be that, like a widower left behind, when the jetpack dream, his lifelong companion, had finally died, Moore himself had nothing left to live for. His daughter offered a less maudlin analysis: "Mom and I always said, 'It was a good thing the Good Lord took him when he did,' because he would not have aged gracefully. He just couldn't."

Following Moore's death, Williams acquired Bell's intellectual property for a sad song and funneled the WR-19 technology into a much more viable and lucrative market—manufacturing Tomahawk missiles. (The company did try to resurrect Jet Belt technology briefly, throwing the kerosene-powered engine into its Williams Aerial Systems Platform, or WASP I and II, which journalist Jaime Wolf described in the *New York Times* as a "250-pound cylindrical tub resembling a flying garbage can," zipping at 60 mph for as long as thirty minutes. Wolf summed up the technology's demise this way: "The WASP II was featured in Jane's All the World's Aircraft as late as the 1984–1985 edition, but again the army, which had financed its development, lost interest. Further attempts by Williams to market it under the name X-Jet failed, and then . . . nothing.")

By all rights, the history of the jetpack should end here. But the genie had a few more tricks up her billowing sleeve. Thousands of spectators had been amazed by Bill Suitor's Disneyland Rocket Belt exhibition during the Christmas holiday of 1965, but only one of them decided to build his own flying machine. That was Nelson Tyler, a tall and rangy Hollywood entrepreneur with energy to burn. He'd been studying aeronautical engineering, but gave it up to follow his dad, a Disney executive, into entertainment. His contribution to the industry was a camera-mount contraption that would allow for smooth tracking shots to be taken by helicopter. "It took me two years to make it, and when I did a little demo reel, it was an instant hit," Tyler said. It was first used on a small-budget sci-fi flick called *Satan's Bug* but has since become a standard piece of moviemaking equipment.

When he saw Suitor blast off, Tyler was struck with another idea for a new product: "I had to have one." He had coveted such a device since he was an eleven-year-old boy, lying across the back bench seat of his dad's '46 Packard, as they wound through Los Angeles's Griffith Park. His father fiddled with the AM dial, the static dissolving when he tuned into a radio play. "It was Buck Rogers," Tyler remembered. "His girlfriend had gone out of the spaceship, and they had gone down to this weird planet and they had landed in this little valley with tall grass. And then the whole valley started to close in like it was a big plant, a man-eating plant, but they escaped with their rocket belt, and that just, like, burned in my mind."

Two decades later, the day after seeing Suitor fly, he took his camera to Disneyland for another demonstration. He photographed the Rocket Belt from every possible angle and brought the stills back to his office in the valley. There he dusted off an old aeronautics textbook and set about re-creating Wendell Moore's invention, selling off a $15,000 sports car to help finance his obsession.

He soon had a brilliant-looking machine—that couldn't lift him over a silkworm. Tyler called the one man he thought might be able to help him; his timing was impeccable. Bill Suitor had quit his job at Bell that day. He caught a three o'clock flight bound for Los Angeles. With Suitor's assistance, Tyler was able to tweak his Rocket Belt knockoff and get it into the air.

When Williams bought out Bell, it had created a giant hole in the futuristic flying business, and Nelson Tyler now soared right through it. In the summer of 1971 he and his now ex-wife, Susan, were the stars of a Canadian Club print ad. A panel of photos depicts Tyler elevating heavenward, the bucolic, snowcapped Tantalus Mountains of British

Columbia rising majestically in the distance behind him. Later, he and Susan enjoy a round of whiskeys with a friend who wears tinted shades and a red bandanna tied at the neck. The copy reads, in part, "All you need is a rocket pack, a pretty assistant—and you're ready to hop your first mountain." Then, down the page a bit, it continues, "All I could think of was . . . next hop Mt. Everest."

It wasn't long before plenty of other offers rolled in. The Super Bowl needed a memorable halftime show. Malcolm Forbes wanted a word with Tyler. So did Soichiro Honda, who delightedly strapped on the Tyler belt, though he opted not to fly. "He had three limos full of guys bowing," Tyler said, still marveling. "Lots of guys bowing."

Sensing a significant opportunity, Tyler convinced Suitor to come out of his grounded retirement and do television stunt work on some shows that today still make sense (*The A-Team, The Six Million Dollar Man*) and others that make less sense (*Gilligan's Island, Newhart*).

For a Pabst Blue Ribbon TV spot, a white-space-suit-clad Rocket Belt ace prepares to zip over desert mountains, a dense cloud of exhaust fired at takeoff. In a classic tough-guy spiel, the gravel-voiced pitchman intones: "You are about to see something very few men have ever tried—to fly without wings, with just a Rocket Belt on their back. Wanna know what it's like? You'll have to try it yourself." Cut to a shot of a dorm-room fridge, stuffed with PBR. "Wanna know about quality in beer? You'll have to taste it yourself. . . ." Cue sand-kicking landing, the pilot relaxing postflight with a cold brew.

The tinkering wizard had created an incredibly successful side business. Tyler's American Flying Belt Company did so well, in fact, that the founder hired another former Bell pilot, Peter Kedzierski, in order to keep up with demand. But during an ill-fated maiden voyage for Tyler, Kedzierski fell twenty feet, crashing down on an Australian aircraft carrier. He was lucky to limp away with severe but not deadly injuries.

Suitor and Kedzierski were sufficiently spooked, and before long both men hung up their flight suits. It seemed Tyler's well of good fortune had run dry. Negotiating a deal to participate in the opening ceremonies of the 1984 Summer Olympics at the Los Angeles Coliseum, he found himself without a pilot he could trust with the biggest moment of his life.

Through friends of his in the local skydiving community, he had been put in touch with a young stuntman named Kinnie Gibson, a brash self-promoter with an outlaw's swagger. Tyler trained Gibson to fly a few jobs for him but was reluctant to use the pilot in the moment of truth. So once more he picked up the phone and called an old friend. "Suitor was so much better," Tyler said. "And I didn't want to exclude him. That pissed Kinnie off."

He was now married (to a Bell exec's daughter, no less) with seven kids and working for the Niagara Falls power company, but Suitor turned the offer into a minivacation, bringing his family out to the Golden State to make a little history. And though he'd flown a Rocket Belt more than a thousand times in his life, the man once nicknamed "Peter Sellers" wanted to be dead sure he didn't screw this one up. So he hit the valley for a refresher course. "Nelson met me at the airport looking like a Beach Boy," Suitor recalled. "We hopped in his Ferrari, and off we went."

To help Suitor get back in the saddle, Tyler loaded the belt into the back of his pickup truck, Suitor riding shotgun. They drove to a secluded spot near the coliseum and parked on a desolate stretch of road. There, twenty years after first flying for Wendell Moore, Suitor once again shimmied into the fiberglass corset to fly high above the redwoods.

On July 28, 1984, Suitor found himself standing atop the highest step of the Los Angeles Coliseum's bleachers, ninety feet above the playing field, on a typically cloudless Southern California afternoon. He was ready to jump. He was the most experienced Rocket Belt pilot in the world—but this was different. This was the opening ceremonies of the

Olympic Games. Nearly one hundred thousand coliseum spectators and another 2.5 billion television viewers around the world were watching. As he waited for his signal, Suitor tried to calm jangling nerves. "I just didn't want to screw up," he told me.

Banishing the thought, he concentrated on the paper sign safety-pinned to his chest that read "Welcome to LA 84." With any luck, it wouldn't tear off midflight. Tyler stood nearby, clutching his 35-millimeter camera with sweat-shined hands. He made a final check of the machine's nitrogen tank—it was pressurized perfectly.

And then Suitor jumped. He swooped out over the field, half-bird, half-man. So many cameras clicked at once that it felt like an unnatural bolt of lightning had struck. "There he is," gasped legendary ABC sports announcer Jim McKay. "Jet Man flying through the stadium, no wires, no tricks, just as you see it."

The flight lasted seventeen seconds and ended with a smooth landing on the turf. Suitor jauntily saluted President Reagan and fell into his wife Cheryl's arms for a congratulatory kiss. "I could have done a little more than 17 seconds, but all that the [ceremony's director] wanted was something to get everyone's attention," Suitor would later tell the *Los Angeles Times*. The next morning he was featured on the front pages of newspapers all over the world. The *Los Angeles Herald* described the scene in a single word, screaming its headline in a World War II–worthy eighteen-point font: WOW! "That's the most famous rocket-belt flight," Tyler said. "For us for sure and apparently for everyone else."

An ordinary man may have been tempted to ride the Olympics success to extreme financial glory. But Tyler is not an ordinary man. Besides, he already had a thriving business. By now he was no longer shooting films himself but was renting the camera mounts out all over the world. Though the coliseum flight had made his Rocket Belt company more popular than ever, he saw it for what it was—the pinnacle. And on top of everything else, in the age of liability it was growing increasingly difficult to secure hydrogen peroxide fuel concentrated to the necessary 90 percent purity. Companies, fearful of accidents, were becoming loath to sell such a volatile material to private adventurers.

Three years after the Olympics, Tyler sold his invention to Tivoli Gardens, the Disneyland of Denmark, for $250,000. The amusement park, in turn, hired Tyler's former pilot, Kinnie Gibson, to perform. By then Gibson had become a busy B-list stuntman, doing recurring work as Chuck Norris's stand-in on *Walker, Texas Ranger* (on which the Rocket Belt once made a cameo), and in films such as *RoboCop 2* and *Police Academy 6: City under Siege.* When Tyler returned to the States, Tivoli had no replacement; it leased Tyler's belt to Kinnie, who quickly formed his own exhibition business. He soon hit the airborne jackpot, flying twenty or so seconds nightly during Michael Jackson's 1989 *Bad* tour in an illusion-creating routine, wherein it appeared it was actually the Gloved One blasting off the stage. Gibson earned a reported $1.4 million for the gig. Over the past two decades, his Powerhouse Productions has put on shows everywhere from Egypt's pyramids to the Brazilian rain forests to Turkmenistan, the Super Bowl, and the 2007 Rose Bowl Parade, charging as much as $18,000 for twenty-five seconds worth of thrills.

That sizable payday has prompted a handful of others to pursue their own Rocket Belt knockoffs. In the mid-'90s, one of Gibson's crew members splintered off with dollar signs dancing in his head, looking to form a new company. But that disastrous venture soon collapsed amid ruthless infighting, bloody battles for control, kidnapping, and, finally, tragically, murder (more on this sordid affair later). Energy-drink entrepreneur Troy Widgery has fared a bit better, and currently has a working rocket belt that he unveils frequently at outdoor events in order to promote his Go Fast! sports beverage.

There are also, as I write this, a handful of middle-aged men— and this level of commitment remains entirely a male province— tinkering away daily in garages, workshops, and their backyards trying to re-create the mercurial magic of Wendell Moore's Rocket Belt. For these singularly driven gentlemen, the welding, lathing, soldering, and bending of parts is a hobby; it's done for their own pleasure, sure, but it is also done to perhaps one day cash in.

They remain undaunted by the sheer difficulty of flying such a device (Bill Suitor has compared flying a Rocket Belt to standing on a beach ball bobbing in the middle of a swimming pool). Or the fact that it will cost literally hundreds of thousands of dollars in parts and fuel to even begin testing the thing on a safety wire. Or even that, if they can somehow actually do what only three others have done before them, in the best case scenario they are talking about flying aloft for about as long as it took you to read this sentence. As mechanical and aerospace engineer Frank Dickman once told *Popular Science,* "Building a rocket belt requires experience in several branches of engineering, [as well as] persistence, courage, serious money and a lot of free time. It's easy only if you're already a rocket scientist."

Whereas the reality of the design and the laws of physics have conspired to restrict real-life jetpack voyages to under half a minute, television, film, Madison Avenue, and the Internet have gleefully shirked off such buzz-killing shackles. The technology continues to be frequently utilized in these visual media.

After Buck Rogers and James Bond, the character perhaps most closely associated with jetpacks is the cool killing machine known as Boba Fett, a bounty hunter from George Lucas's original *Star Wars* trilogy. Film audiences were first introduced to Boba Fett, laconic behind a chipped green space helmet, Wookiee scalps dangling from epaulets, in 1980's *Empire Strikes Back.* (Fett had made an animated appearance in the televised *Star Wars Holiday Special* in 1978.) But it wasn't until 1983's *Return of the Jedi* that we saw him fly. I can still recall my own glorious epiphany: That missile launcher on his back? Yeah, it's actually a jetpack!

Boba Fett remains one of the most popular characters in the entire *Star Wars* galaxy, particularly with guys who, like me, came of age during the Force era. Web sites scrutinizing and analyzing Fett's biography, motivations, attitude, and, especially, his gear continue to proliferate. On the site Boba Fett Multimedia Page (www.bobafettmp.com), you can visit the Dented Helmet Forum for answers to all things Fett. As of the spring of 2008, a posting in the forum "Official Lisa Fett's Jet Pack Tutorial" had been read by 10,439 Fett heads. Looking to build your own flawless bounty-hunting costume? TK409.com (if you have to ask about the name, you probably shouldn't go there) can help.

Over at the mother ship, Bobafett.com, some industrious geek has taken the time to painstakingly annotate a photograph of Fett's weaponry, armor, and outfit. There are two entries on his jetpack, found just below information on gloves and wrist gauntlets. They tell us how in the case of the Mitrinomon Z-6 Jet Pack the "fuel tank holds enough fuel for one minute of continuous operation (20 three-second blasts)." About the Rocket Thruster Fuel Tanks, we learn how "each three-second blast moves Fett up to 100 meters horizontally or 70 meters vertically. Top speed of 145 kilometers per hour; maximum range of two kilometers." Good to know.

Not long ago a nerd-core anthem written by the thirty-three-year-old actor, rapper, and comedian known as MC Chris, "Fett's 'Vette," circulated as an MP3 on the Web. It is a relentlessly spunky, rat-a-tat rap containing the thrilling chorus:

My backpack's got jets
Well, I'm Boba the Fett
Well, I bounty hunt for Jabba Hutt
To finance my 'Vette

It's hard to say what exactly accounts for the ever increasing levels of Fett-ian devotion; George Lucas himself has said in interviews that he has no idea why the character caught on so well. (It's additionally perplexing when one considers that Fett has extremely limited screen time in both *Empire* and its follow-up, *Return of the Jedi.*) Perhaps it is his renegade spirit. Maybe it's his uniform—the dusty green cape, camouflage cargo pants, and tinted helmet are an undeniably cool look. Or maybe it's his mechanized robot voice.

Joe Johnston, a film director who broke into the business at Lucas's Bay Area effects mecca, Industrial Light and Magic, and helped design Boba Fett, framed the appeal in a simpler way when he asked, "What kid wouldn't want a jetpack? There's this universal fantasy of—what if you had one of these things? It's such a dream to be able to fly it."

Interestingly, Johnston told me, Fett's origin was the result of a spontaneous decision. Early on Lucas had in mind an army of ur–Storm Troopers called Super Troopers, who wore similar all-white uniforms. But by the time he began developing a bounty-hunter character, the Super Trooper idea had been abandoned, so the director told Johnston to take one of the unused uniforms and turn it into something an outlaw might wear. "Since Boba Fett had built this suit himself, had found it at a yard sale or flea market, George asked us to paint it up like the suit of a renegade, who put it together himself with a jetpack on the back," said Johnston.

That jetpack, an unencumbered tourist's satchel in design compared to Wendell Moore's rucksack, was made of vacuum-formed styrene attached to a camping-gear frame. It was weighted with two-by-fours so as to not flop around. The exhaust flames were worked up in postproduction.

The one and only time Boba Fett flew it onscreen, things don't go so well. The scene takes place above the toothsome Sarlacc Pit, which looks a bit like an anus, if an anus had giant space fangs and was reshaped during a collaboration between David Cronenberg and William S. Burroughs. Jabba the Hutt is hovering on his sail barge, preparing to sacrifice our heroes. A scuffle breaks out between Han, Luke, Chew-bacca, and Lando Calrissian on one side and Jabba's flunkies on the other. A jetpack-enhanced Fett zips across the pit to a smaller vessel where the fight is in full bloom. His technique is pure superhero, hands-over-head, as if diving into a pool. So far, so good. But just as he's about to shoot Luke with his wrist blaster, Han, who is blind at this point in the narrative due to your standard cryogenic deep-freeze mishap, accidentally nudges him in the back with a space spear (or "vibro-axe," if you will), activating his 'pack and launching Fett wildly out of control, smack into the sail barge. The bounty hunter slithers off the ship like a dead fly on a windshield and falls into the sandy maw of the Sarlacc. The death is gruesome but kitschy. This is the first pop culture moment when a jetpack neither belongs to a hero nor facili-tates heroism; to the contrary, countless adolescents must now con-sider the idea that jetpacks can be bad for your health.

That kicky and doomed portrayal continues to resonate in con-temporary pop culture representations of jetpacks. In a 2005 episode of the hit Fox adult cartoon *The Family Guy* a few years back, the title character's jetpack malfunctions, ramming his head repeatedly into the doorjamb like a human jackhammer. In a 2000 television ad for Budget Rental Cars, a conference room full of employees is brain-storming to see if there is a better way than via shuttle bus to get cus-tomers to their cars. One guy suggests "jet-propulsion packs." Cut to the hapless renter's fiery encounter with a telephone wire. They stick with the shuttle.

Two of the most popular comedians on television, Jon Stewart and Stephen Colbert, have made jetpacks recurring punch lines. Return-ing from a commercial break during one episode of his fake news pro-gram, *The Daily Show,* Stewart turned to his guest Bill Gates and asked, apropos of nothing, "When are we going to get jetpacks?" His audience of young wiseasses boomed with laughter, clearly in on the joke. Gates's answer: "We're not working on that one." (Gates, or at least a cartoon version of Gates, did, however, don a 'pack on another *Family Guy* episode.)

Soon after assuming the host's chair on his own news parody program, *The Colbert Report, Daily Show* alum Colbert delivered this riff, steeped in his signature faux gravitas, cocked eyebrow firmly in place: "For years, they've been telling us we're going to have little wrist televisions, one-man rockets, and whole meals in little pills—hopefully, this year, we can do it." To which I say, Amen, brotherman! I mean, personally, you can keep the little wrist televisions and meals in a pill. But one-man rockets? Jetpacks? Hell. Yeah.

Some celebrities still believe in the pure alchemy of the machine, its ability to turn reality into fantasy. When Powerhouse performed on David Letterman's *Late Show* in 1994, the usually wry host flashed a rare expression of unironic and unbridled enthusiasm when he declared, "I'm the celebrity—I should have a jetpack!" Though it is unclear if he used a stunt double (calls for confirmation were not returned), P. Diddy appeared in *People* magazine photographs in 2005 riding a jetpack to a press conference for MTV's Video Music Awards, which he hosted that year.

The post-*Jedi* era also gave rise to the most jetpack-centric film narrative to date. In 1991 Disney made a movie version of an intensely popular graphic novel homage to pulpy 1930s art deco Los Angeles. *The Rocketeer* tells the heroic story of one man's struggle to defeat the Nazis, win the war, and get the girl. He couldn't have done any of it, however, if he hadn't stumbled across the stolen flying machine designed by legendary recluse and airman Howard Hughes. After finding the X3 RocketPack amid the workshop detritus of a kindly scientist, Cliff Seacord is transformed into the titular superhero. His helmet may make him "look like a hood ornament," in the words of the scientist, but his 'pack is all sleek coppery brilliance.

Though the film was neither a massive box office nor a critical success, it did provoke an impressively loyal and passionate cult following. Joe Johnston, the Boba Fett designer, directed the movie, thus establishing himself as a Hollywood darling to jetpack junkies of all stripes. He told me that he still gets frequent letters from a few hardcore fans, effusively praising the film. One such guy, a Canadian named Brad Brown, with maniacal attention to details, constructed his own Rocketeer outfit—brown leather jacket, khaki lion-fighter pants, jetpack, hood-ornament helmet—and redesigned his living room to look just like the film's Bulldog Café, in the shape of, you guessed it, a bulldog. Johnston seemed more bemused than alarmed by such intensity of feeling. "The idea of flight is so alluring," he reasoned. "You can get obsessed."

In a sense, it's easy to see why fans would react with such intensity. As one long glide down the glittering California coast makes plain, *The Rocketeer*'s equipment enables flights much longer than twenty-one seconds.

Over the past few years there've been several other big-screen blastoffs—again, some make sense to the narrative (science-fiction-y fare like *Minority Report* and *Sky Captain and the World of Tomorrow*), whereas others are bizarrely incongruous (*Jingle All the Way*, anyone?).

Unsurprisingly, online and in video games, where geeks and garagistas alike thrive, the jetpack has done well. Games as disparate as *Grand Theft Auto: San Andreas, Donkey Kong*, something called *Dangerous Dave, Ork, Scram*, a *Star Wars* quartet, and many other titles all incorporate the technology into their storylines. Almost exclusively, jetpacks are portrayed as still cool and cutting edge in the gaming universe.

The same can't be said of the Internet, where Web sites either dedicated to the technology or appropriating it in their design and name strike a balance between mocking languishing jetpack development and bemoaning the sadness of that fact. Two young admen each operate their own jetpack-themed blogs. One is called "They Promised Us Jetpacks and We Got Blogs"; the other is, simply, "Where's My Jetpack?"

At the top of the latter blog's home page is an image of a clean-cut guy, a '50s-style space pioneer, jaunty in a sparkly, fire-retardant suit. You can practically smell the Aqua Velva wafting off the screen. One problem, though, is that he's not wearing a jetpack! Intrigued nonetheless, I instant-messaged with the site's operator, a man whose screen name is jetpackjockey.

Me: so . . . you're really into jetpacks. so am i.
Two minutes later, jetpackjockey responded: it's more of a joke about what they said would be "in the future"
Me: and nothing says future like a jetpack, right?
jetpackjockey: absolutely
Me: what is the allure?
jetpackjockey: I think there is a general disillusionment among people who grew up with a futuristic fantasy—a feeling like we got ripped off almost—nothing's changed and the world is going to hell faster than ever.

It's a notion that Mark Wells, the Alabama NASA researcher and friend to Jet Vest inventor Thomas Moore, once conveyed to me. When I asked Mark why so many boomers were possessed by the jetpack's voodoo, he said, "My opinion is that it represents totally free, uninhibited flight. In an airplane or a helicopter, you are at the mercy of cumbersome aerodynamics. This is totally free flight and totally controllable, a level of control you don't have in a plane or helicopter." Then he paused and added, "For a lot of us, the future once looked very exciting—so it's a nostalgia for that lost future. Now the future looks more scary than anything else."

It isn't just boomers reminiscing for this lost future. Daniel Wilson is a thirty-year-old roboticist and author. His 2005 book, *How to Survive a Robot Uprising*, part kitsch, part real science, looked at, well, how to survive a robot uprising. In 2007 he published an examination of thirty once-promising technological wonders—flying cars, underwater hotels, teleportation, universal translators—looking at whatever happened to such utopian pursuits and if we'll ever see them outside of an episode of *Star Trek*. He called the book *Where's My Jetpack?* I asked him about the title. "It's about this notion that we're in the future now," Wilson said, noting that once the calendar flipped past the year 2000, he observed a shift in the way his peers perceived technology like jetpacks and flying cars. Where once they may have joked lightheartedly that the scientific promise has not been delivered, now they are prone to say, "No, really, where's my jetpack?" "My cohorts are all twenty-five to thirty-five years old," Wilson continued. "And maybe this isn't true of a lot of twenty-five to thirty-five year olds, but they are all trying to relive their childhoods and they all want to know where their jetpack is." Blame Boba Fett.

Nostalgia also plays a part in the decision of advertisers to push the technology. From golf equipment to the DVD delivery service NetFlix to the computer troubleshooting outfit the Geek Squad, jetpacks continue to occupy a meaningful corner of the ad business.

My friend Greg Mills has worked in advertising for twelve years. He is now a senior copywriter for McCann Erickson in San Francisco. "Marketers, specifically the people who make ads, want to instill a sense of inclusion with their target audience," Greg told me. "A very lazy way to do this is to pick on the outmoded aesthetics of the past. If you throw up a jetpack or a mullet or whatever, you can point to that dated icon and say, 'Ha-ha! Remember that crap? God, we all sucked back then. But we are presently with it, and we can laugh about this together.'"

This is interesting but does little to explain one gloriously twisted, jetpack-fueled online rant so crammed with pungent prose, so heartfelt and heartsick and from-the-gut intense, that, going forward, I might have little choice but to appropriate it as my own manifesto of sorts. Ladies and gentlemen, from New York City's "Best of Craigslist" files, I present you with the infinite wisdom of some long-suffering anonymous soul who captures and articulates in amazingly unsubtle terms the thrust of what I am now beginning to see as a very real Where's My Jet-pack? movement. This guy couldn't give a damn about mullets, I'm guessing, but a jetpack? That's serious business. I probably don't really need to say this, but any all-caps are all his:

WHERE'S MY PERSONAL JET PACK?!

DATE: 2003-12-31, 1:26PM EST

I am sick and tired of waiting for the personal jet packs we were promised when we were growing up in the 60's! All those black and white films I saw in school of what the future would be like IN-CLUDED PERSONAL JET PACKS! It's 40 fucking years later and even our cars still require that WE STEER THEM (also a lie from those films about the future). The only "big" invention of the future we've gotten so far is the fucking SEGWAY. It's a SCOOTER, dude! We had scooters in the 60's! "Ooh! You can stand on it!" You can stand on a fucking scooter too, ass! "Yeah, but, it won't fall over no matter what!" A scooter is only THREE INCHES FROM THE GROUND, dumbass. FALLING over was never, like, my primary concern. I am sick and tired of sitting in traffic FOR HOURS. I have been living with it for twenty years. In 1980, TWENTY years after the promise of a personal jet pack that would fly you anywhere in a minute, I was still sitting in bumper to bumper traffic. But, I thought, just hang in there, SOON, they will have personal jet packs, FOR SURE by the year 1992. Now here it is 2003! It's the fucking twenty-first century dude! And to the inventor of the Segway, and Bill Gates who I'M SURE had some-thing to do with it, because he's a dick! You should have your bare asses WHIPPED for trying to create a big stir about the SEGWAY be-fore it came out by saying shit like: IT WILL CHANGE THE WAY WE LIVE. That LED ME TO BELIEVE: PERSONAL JET PACK!!!! NEVER NEVER NEVER CLAIM that AN ELECTRIC SCOOTER will change SHIT!!!! IT'S A FUCKING SCOOTER! When we can FLY TO WORK,

THAT WILL CHANGE THE WAY I LIVE. WHEN WE CAN DRINK BEER AND EAT CANDY AND LOSE WEIGHT, THAT WILL CHANGE THE WAY WE LIVE. WHEN I CAN GET A FUCKING HUMAN BEING ON THE GODDAMN TELEPHONE AT THE FUCK-ING DMV/PHARMACY/CABLE COMPANY, THAT WILL CHANGE THE WAY I LIVE.—NOT A SCOOTER! And stop trying to make it seem cool to have a Segway. IT'S NOT. Dick Tracy (written in the fucking 1930's) had a FLYING Jet pack, not a scooter, buttwipe. It FLEW! In 1930! That's before we had MICROWAVE OVENS and peni-cillin, and organ transplants. 73 years ago! I want my personal jet pack! WHERE IS MY PERSONAL JET PACK! Everyone, if you email congress and tell them to quit throwing away that dumb ass money on science and funnel it to flying jet pack research, WE CAN MAKE THIS HAPPEN!!!! PERSONAL JET PACK!!!!!!!

There is certainly much to admire here ("our cars still require that WE STEER THEM," e-mailing Congress, and so forth), but one bit that particularly intrigued me was the time and date, early afternoon on New Year's Eve. Dude was home, and one can't help but imagine he was alone, facing the prospect of bidding adieu to the futuristic-sounding year of 2003 and ushering in the even more futuristic-sounding 2004. And he was to do it sans jetpack. Disillusioned? I should say so. Inspired? Damn straight.

Duly inspired myself and feeling well steeped in jetpack mythology, I am now ready to forge ahead, to find my own flying machine. And, as luck would have it, an event that promises to be enormously helpful to my pursuit fast approaches.

CHAPTER
3

Shuffling Off to Buffalo; or, The Cruel Truth about Lift over Drag

Once you have tasted flight, you will forever walk the earth with your eyes turned skyward, for there you have been, and there you will always long to return.

—**Leonardo da Vinci**

An old Silicon Valley proverb says, "If a thing exists, there is a Yahoo! group for it." That's a rough translation from ancient Geek, but you get the idea.

The Yahoo! Rocketbelt Group was founded in 2003 by Peter Gijsberts, a divorced father of one, who manages a (ground) transportation company and lives in the southern part of the Netherlands that Vincent Van Gogh found irresistible. Peter's related Web site, rocketbelt.nl, is a vast clearinghouse dedicated to all that is cool about the Bell invention of the 1950s and '60s. It is highly recommended for anyone who wants to get lost in dreams of flying for an afternoon or ten.

After high school, Peter studied safety and security management but could never shake his significant fantasies of flight. He launched the Yahoo! group after falling under the spell of Wendell Moore's genius, and, as he told me in an e-mail, he hoped to "get more people enthusiastic, to share knowledge, to help others, to get more info on the subject, to make friends forever." And as he once told a reporter, "Soon, it became an obsession."

Today, the group has about 140 members who gather electronically from all over the world—Australia, Spain, Germany, England, New Jersey—to admire, praise, relive, and in a handful of cases even attempt to rebuild Bell Aerospace's 1960s Rocket Belt. For those ambitious, hard-core fans looking to get themselves airborne, the countdown to liftoff has been long (forty years in one case), dangerous (as in murder), and expensive—it is, apparently, possible to spend hundreds of thousands of dollars outfitting your home workshop with new giant lathes, enormous welding machines, a home mill, and many other assorted power tools.

Former Bell Rocket Belt pilots Bill Suitor and Hal Graham are members of the Yahoo! group, though they don't often chime in. Here those two are anomalous superstars. The more typical member is your average, domesticated middle-aged dude with a bit of engineering acumen and a serious hard-on for *Lost in Space*. They have online handles like "rktman8888" and "rocketbelts" and, naturally, "jetpack."

"Throttle Valve Operation" is an example of a classic subject header in the forum. It was in an e-mailed conversation thread bearing that subject name that a group member once wrote: "If I understand the valve operation correctly (and I'm fairly certain I've got it nailed down, or at least something that will work as well), the part that came out and stuck in the wall was the plunger. The plunger is the portion that performs vernier throttle control, while the spool provides gross flow control from 0 to ~70%. This becomes apparent when you look at the flow test graphs I posted in the files section from the SRLD TRESCOM phase1 report."

One needn't understand a single word of that to know that Rocketbelt Group members are fanatical about the throttle valve; it is barely overstating the case to say that the throttle valve is a Rocket Belt obsessive's Holy Grail. A few years ago, one member came across the original Bell Rocket Belt plans and has sold or bartered the hot property to several other group members. Some of the builders in the forum have spent thousands of dollars and years of their lives trying to perfect their own valve technology.

This makes sense when you consider that if you actually plan to build and fly one of these things, a well-made valve could be all that stands between you and extreme bodily harm. If the valve, which looks sort of like a small metallic bike pump minus the hose, doesn't very precisely push 24.5 ounces of highly pressurized hydrogen peroxide into your catalyst chamber *every second,* you won't stay aloft for very long. Seventy pounds may not seem like much weight on your back until it is forcing you facedown into the cold, cold ground at 60 mph. And if the valve jams open, by the end of that bumpy adventure you'd look like a G.I. Joe doll run through a wood chipper.

The forum is generally a place of goodwill—a community of mostly like-minded boomers who've stuck their flag in a piece of technology that summons in them warm thoughts about a past brimming with potential and the possibility of an extended adolescence. Occasionally, however, virtual brawls break out among competitive builders who want recognition for how much work they've poured into their projects. There is a premium placed on being first—at anything. The First American Amateur to Fly on a Tether. The First European to Go Off Tether. The First Mexican to Build Four Rocket Belts. This is probably not that surprising, considering most of the group members came of age at the height of the space race, a time of so many firsts.

A few months after joining the group, I open my e-mail to find a message that prompts both great curiosity and more than a dash of trepidation. It seems that Peter is making plans for the first ever

"International Rocketbelt Convention." Members are invited for a week-end in Niagara Falls, just a short drive—or currently unimaginably long jetpack flight—from where Harold Graham first flew free. There the local air and space museum will host former Bell engineers, pilots, and their families, as well as enthusiasts, tinkerers, and builders from all over the world. It is to be a celebration of Wendell Moore and his lifework. It's a weekend salute to diligence, determination, and protective eye-wear. As well as dreams, obsessions, and an astounding passion for leaping heavenward. There will be lectures, presentations, symposia, and PowerPoint. Oh, and weather permitting, there will also be a Rocket Belt demo or two courtesy of Troy Widgery and his Go Fast! energy drink's promotional department. The idea that I might actually, finally, see a man fly through the air while not lunging for a foul ball over the upper-deck railing is reason enough to make the trip.

And while I'm a little nervous about meeting my fellow members in the flesh—what can I possibly say about the throttle valve that hasn't already been said a million times already?—overall I think this is a very good thing. In the nearly fifty years since Moore first showcased his machine's promise, there have been scarce improvements upon his original design. Perhaps this convention will finally jump-start the next wave of inventors and inspire them to hop off their Segways for a minute to focus on something truly important.

But first I need a crew. The late-September convention is still a few weeks away, so that gives me plenty of time to round up a couple of hotshots to help infiltrate deep into techno-nerd territory and emerge unscathed with new and vital information about the contemporary state of the jetpack. Also, I could use a photographer.

It doesn't take long to draft my dream team. Jofie has already been identified here as among the greater jetpack junkies I know. Physically, Jofie calls to mind Gay Talese's description of George Plimpton's "long, skinny limbs, a small head, bright blue eyes, and a delicate, fine-tipped nose." He is kind, quick-witted, and charming, so I know he'd have those space-age pioneers eating out of his hand.

Next in command will be Joanna, a friend I'd met at school in Santa Cruz. Joanna is a fantastic photographer and an endlessly interesting, self-identifying dilettante. She will be great company for twenty hours of driving over three days. For one thing, but certainly not the only one, she makes a mean mixed CD. With her reddish bob and wardrobe of vintage skirts, she reminds me of an indie rock star as imagined by Charles Schultz. And let's be honest, it will be good to have a woman around to charm a crowd that promises to be both very male and very likely a bit socially awkward. This convention is, after all, a kind of humble stepchild to Dragoncon, the enormous annual sci-fi geek fest. Joanna, I am confident, will very capably document the proceedings while also catching any idle gossip I might miss.

With the team in place, there's nothing to do but wait. Well, wait and sweatily bounce my newborn daughter on my shoulder, hoping against hope that she might finally, mercifully, doze for a minute or two. Daphne Byrd was born almost exactly two months before the convention.

When late September rolls around and it's time for me to head north, I know I'm going to miss Catherine and the girls terribly. Truthfully, it's more Oona I will miss than Daphne. The Byrd is a beautiful, red, roly glob of flesh, and I love her very much. But we don't have much to talk about, at least not yet. Oona, on the other hand, is fast becoming one of the more impressive conversationalists I know, which I honestly think says more about this particular two and a half year old than it does about my friends.

We talk a lot about the fact that she lives in Booklyn, Noo Yawk; loves cheese, milk, and other dairy products; and very much enjoys going "superfast" when riding in her little plastic seat on the back of my bike. Oona has already mastered Silly Walks and Funny Faces, so her comedic chops are really getting there. Which is great, except when it comes to leaving to go in search of a jetpack. With her long, easily tangled surfer-girl blonde hair and enormous, round blue eyes, I often found myself looking at her in the days leading up to my departure, wondering how someone like me could've helped produce something so objectively beautiful.

And Catherine? Here's all you need to know about Catherine to understand what level of superstar we're dealing with: she is letting me desert her with two babies while I go looking for a machine that, as far as I knew then, didn't really exist. Her only condition is that I can't actually fly a jetpack. And if I do, I can't die. The kids need me, and, besides, I'm between life insurance policies.

On the morning before the convention is to begin, I pack my overnight bag with underwear, socks, notebooks, and our video camera. I kiss my girls good-bye and feel as though I've been punched in the heart. This is just the first of many difficult good-byes over the next several weeks, and every time I leave I'll be wrenched anew, thinking that I've set myself a potentially dangerous course and cannot wait to return in one piece. I'm going jetpack hunting; I have no idea where it might lead.

Getting out of Brooklyn turns out to be the trickiest part of the drive. While making the rounds to pick up Jofie and Joanna, I am twice forced to swerve suddenly to avoid a road-enraged gunner. It doesn't help that I haven't yet had my very necessary morning coffee and I've left my apartment, bleary-eyed, before seven to try to beat the traffic.

But soon enough we are through the Holland Tunnel and then into the part of northwestern New Jersey that is very mountainous and green, with wide pastures on either side of the road. We are cruising, making good time. Jofie is in good form, completely fired up about what is to come, though none of us has any idea what might occur or if there'll be five attendees or five hundred. I mean, who even knows about this stuff, let alone cares enough to schlep beyond Buffalo to see it?

"All I know is, I'll pay a thousand dollars to fly a jetpack," Jofie says, leaning forward and scanning the highway. He works in the publishing industry, and so I'm pretty sure he doesn't have a thousand bucks to throw around. Still, I believe him.

After a quick stop in Syracuse for lunch at a semilegendary barbecue joint, we make our final push into Buffalo.

Like a lot of American cities, Buffalo is a once proud and thriving metropolis. Located on or near the banks of several major waterways—the Erie Canal, Niagara Falls, and Lakes Erie and Ontario—for much of the nineteenth and early twentieth centuries, Buffalo was a powerhouse in the booming steel and grain industries. Western-bound pioneers happily used Buffalo as a stopover, before continuing their journey by lake or train.

But when a new waterway, the St. Lawrence Seaway, opened in 1957, it allowed ships to bypass Buffalo's port, and the painful economic effects were almost immediately felt. Major company closings and, later, the extreme suburban exodus that took place all over the country caused Buffalo's population to fall steadily every year, from nearly 580,000 in 1950 to less than 280,000 in 2005. In recent years, the town has begun to reinvent itself as a center for cutting-edge biology research in fields with sci-fi-sounding names like bioinformatics and genome assembly.

Buffalo is very much a college town, with about twenty higher-ed facilities in the metropolitan area, bringing some 40,000 students together. Which, in fact, is why my crew and I are staying there, a half-hour drive from Niagara Falls and the convention headquarters.

An old Baltimore friend named Peter had, several years ago, taken a teaching position at one of the local universities and is now charged with making text-message-crazed American-lit students care about Walt Whitman.

But if anyone can do it, Peter can. Growing up, he played drums in and was the spiritual engine of a regionally famous rock band that sounded like Echo and the Bunnymen with a Ballmore accent. Peter had great enthusiasm for weekend fun and a wit by turns sweet and sardonic. The last time I'd seen him perform was about six years ago— by then he was singing in a Grateful Dead cover band at a sailors' bar near the Inner Harbor. Peter has, naturally, become a loving husband, a diligent professor, and a prolific, talented, goateed poet. He and his wife own a lovely, low-slung ranch house with broad front windows and a backyard built for barbecues on a wide, leafy suburban street in Buffalo. When we finally pull up in front, Jofie nearly jumps out of his seat. "We're staying in the Mafia house? Awesome!" And it is awesome— cozy yet 1950s kitschy in all the best ways.

Peter is reading as part of a lineup of poets that night on campus. We round up one other old friend who is coincidentally now a professor at a different local school, stop off for a couple beers, and then hit the reading.

It has been many years since I've heard real poems read aloud. At first it's all I can do to even look at the stage. It is too bracing, too raw and immediate, to hear these hairy, fleshy, disheveled maniacs stand at the front of the barely filled auditorium and, with zero visible self-consciousness, emote. Had they been stripped naked and adorned with nipple clamps, they'd appear less exposed and vulnerable. The poets cram each stanza with sensual, messy, oozing ideas about emotions, relationships, and philosophy. It's pornography of the soul. I feel like a cynical, big-city jackass for finding its visceral qualities so unsavory.

Thankfully, though, when Peter takes to the stage, his first poem name-checks Led Zeppelin. With that pop cultural prick, he punctures the suffocating high-earnestness, and I think I feel the whole room relax a little. Or maybe it's just me.

Afterward, we have a drink or two more than we probably should over dinner and then call it a night. My team needs rest. Tomorrow, the future could finally be ours.

While Buffalo has begun to transform into a high-tech research center, Niagara Falls, New York, most certainly has not. In harnessing the falls' immense power (on average, four million cubic feet of water crash across its massive rocks every minute), the area was, in the early twentieth century, a fast-growing industrial player in the steel, chemical, and manufacturing sectors. But then familiar afflictions struck in the forms of corrupt government, cheap outsourced labor, and the subsequent closing of industrial plants.

Lately, the city has begun to focus on the one export it possesses that still holds any currency in the rest of the world: tourism. The falls will attract some twenty million visitors this year to its thundering banks, and in 2004 the Seneca Nation took over the land previously home to the city's civic center and erected a twenty-six-floor, gaudily glittering casino that dominates the view for miles around. It might very well be the only shiny thing left in Niagara Falls.

Entering via Buffalo, we drive over a narrow bridge and dip down into the city. It is drizzling rain from a flat, white sky, throwing the convention's scheduled rocket-belt demonstration into doubt. Through the soggy, squeaking windshield wipers, I can see a long road lined with small, ramshackle houses, many with boarded-up windows, slouching atop hoary, neglected yards. These sad-looking plots are occasionally interrupted by the sort of once kitschy, poorly aging roadside motels that very well could have hosted the protagonist of Nabokov's *Lolita*, as he took his young charge on their doomed holiday drive.

One holdover from the city's glory days is the lingering presence of power plants owned by such companies as DuPont, DOW Chemical, and Occidental. Soon we are driving under an eerie thatched roof of electrical wires stretching in all directions from industrial campuses taking up block after block of depressed real estate. As far as I can see in every direction, above us hangs a vast network of wiring strung from a warren of minimalist steel towers.

"This is so weird," Joanna whispers from the backseat.

"I can feel the hair on my arms tingling," marvels Jofie.

I'm not sure if it's just the power of suggestion, but at that moment, I, too, feel tingly and not in a good way. It seems reasonable to assume that a toxic accident might occur at any moment—propelling us into a real-life Buck Rogers comic. Actually, it strikes me that if I really am going to find a jetpack, it will very likely be in a place like this—bleach-white sky, desolate roadway, city reduced to nothing but tangled hardware. A place that could've been dreamed up by a cold war–era science fiction writer.

Finally, we emerge from under the roof of wires and come to a traffic light near the museum. To my left sits a dumpy motel called the Rodeway Inn. An image flashes in my mind of the day the balding, mustached owners decide to go with that spelling, Rodeway. The meeting, the handshakes, the backslaps. Too bad about that. To the right I see a three-foot-high weather-battered sign adorned with three layers of faded American flag bunting: "Welcome to a Wonder of the World." Behind the sign, the Seneca Niagara Casino—all 113,000 square feet of its blinking, hysterical, empty promise—rises like a gambler's Death Star—well, like a Death Star, if the Death Star had a gaudy rainbow tribal light design running the length of it.

I'm getting a little depressed. This is not the Niagara Falls of 1950s honeymoon dreams.

"Think the rocket-belt guys are still fired up?" I ask my crew.

"Fuck, yeah," Jofie answers. "This is the first annual." The car goes quiet.

Soon we roll past the Niagara Aerospace Museum—a modest two-floor structure the color of drying mud—and park in the back lot. The drizzle is lightening.

At the front desk, an aging hippie who looks drawn by Matt Groening alerts us to the news that the morning buffet is almost gone. "Go! Hurry! Get some breakfast!" He is, apparently, more excited about cold eggs than jetpacks.

We speed-walk through a ground-floor permanent exhibit in a room barely bigger than a helicopter-landing pad, past a long glass case demonstrating the history of flight, from balloons to ballistics. At a World War II–era bomber, we hang a left and take the escalator to the second floor. There, among the buffet ruins, about a dozen pasty nerds shuffle around the registration table, smiling awkwardly at one other. I count three women, including Joanna.

I don't recognize any of them as the Bell Rocket Belt pilots whose pictures I'd been obsessively ogling online for weeks. At first blush, it could be a convention for anything, really—cat enthusiasts or model-train nuts. These are the anonymous, doughy faces of obsession. Were you to pass any of these people on the street, you'd have absolutely no idea about their heart-wrenchingly beautiful dream, their desire to build with their own hands a shiny metal contraption that in the best possible scenario will lift them off the ground by their armpits, like a twisted bird, for a wingless, breathless twenty-two-second orgasm in the air.

And then I notice an older fellow who is either Harold Graham, the first Bell Rocket Belt pilot to fly without tether, or some homeless guy looking to score a free meal. The reasons I think he may be homeless have mostly to do with his outfit, his black scuffed loafers over bulging white athletic socks, sagging shorts, and a stained white T-shirt. Oh, and he is holding a plastic lawn-size trash bag that may or may not contain a dead raccoon. And he is mumbling to himself. The reasons I think he might be Hal Graham include his spaceman's closely cropped haircut and that familiar-from-pictures sharply jutting, gravity-defying nose and chin—his face the slightly fossilized relic of a 1950s futurist. He is also wearing a mesh baseball hat on which someone has scribbled in black Magic Marker a single word: "Hal."

The convention schedule mentions the possibility of a flight with Hal in his personal Cessna, which he's piloted with his girlfriend from their home in Crab Orchard, Tennessee. I am still not sure if the man in front of me now is Harold Graham, but there is no way I am going flying with *that* dude!

Jofie and Joanna corral a vacant plastic picnic bench and sit down with cold coffee and a packet of literature on the history of the Rocket Belt. I go looking for the bathroom. Just past a few Plexiglas displays on the triumph of imagination over gravity, I turn a corner and find the men's room—and something else, too.

In a larger back room about the size of a high school gym, quietly resting on metal easels, sit what look like—could it be?—five gleaming rocket belts and a rocket chair, too. This is the closest I've ever been to a real-life jetpack. The handlebars poke forward three feet like skinny robot arms. The aluminum nozzles are positioned above fuel tanks on the back, creating something that looks like the world's coolest fire extinguisher, the carbon fiber corset offering the hottest hug you could ever want. After dreaming for so long about what it might feel like to leap into the air and then keep going up and up, it is a shock to suddenly be facing a roomful of machines that could, potentially, enable me to do just that. After all, I'd just been looking for a place to pee.

The room's periphery has been decorated with Bell memorabilia through the ages. On a pair of folding tables, cardboard collages depict several pilots midlaunch. There are snapshots of one pirouetting over Disney's castle, stills from *Thunderball* and *Lost in Space*, cartoons of midair soldiers, knees flexed, engines roaring. A wall is dedicated to Bell's lunar equipment—particularly its Lunar Escape Astronaut Pogo, or LEAP, a kind of flying platform that Wendell Moore had helped design. Looking like the unholy offspring of a garbage can and a metallic spider, the LEAP was ultimately rejected by NASA in favor of the safer, more durable moon buggy. In front of the LEAP display a grinning mannequin in a New Wave orange jumpsuit, white crash helmet, and sunglasses has been erected. Strapped to his back is what appears to be most of an original Bell belt; he is quietly stuck to the ground.

Near the crudest of the six devices forming the glittering semi-circle, the only one conjuring thoughts of science fairs, papier-mâché, and aluminum foil, stands a kind-looking man in a denim shirt, navy tie, khakis, hiking boots, and thick glasses. The high-wattage bulb of a television camera illuminates him. A smiling brunette holds a microphone toward his face. A crowd of ten or so paunchy, balding men stands behind her, anxiously shifting their weight from one foot to another.

I inch closer.

"If you were to just strap a jet engine with a single compressor in it to you, the spinning action of the compressor would create gyroscopics," the man is saying. Who is he? He looks familiar.

"So when you lift off, like a helicopter without a tail rotor, the gyroscopics will start you spinning, just the mass of the engine." He has a pleasant, folksy way of describing the inner workings of a Jet Belt, that's for sure.

"So Sam came up with the grill-shaft engine where the inner shaft was half the size of the outer; it spun at twice the speed of the outer one, and they canceled each other's gyroscopics out. So it was like a flying gyro really, just as stable and solid as a rock."

Sam? Aha—Sam Williams, whose company bought Bell's Jet Belt plans after Wendell Moore died. That means this guy must be . . . Bill Suitor! Or Mr. Jetpack, as I'd seen him ID'd online.

I've interviewed Magic Johnson and not been flustered. I've passed within three feet of Yoko Ono on a Manhattan street corner and felt barely a thrill. So why is my heart now singing, the reverb knocking my ribs? Because this, this man is a legend. And the legend is pointing to parts of the old Bell Rocket Belt, describing what is becoming a familiar process, even to my scientifically challenged self.

"The Rocket Belt uses 90 percent–strength hydrogen peroxide; it's water with an extra molecule of oxygen, H_2O_2, in this little chamber up here." Suitor touches a part shaped like a kerosene camping candle.

"This is actually the rocket motor. There's a silver catalyst—screens with silver and other coatings on it. When the peroxide is sprayed into there and hits the silver catalyst, the peroxide decomposes. That extra molecule of oxygen is released. Now you have water and oxygen, through this chemical reaction. At 90 percent strength, hydrogen peroxide's reaction temperature is 1,388 degrees Fahrenheit in two-tenths of a millisecond—it's instant heat. Now you have water and heat, so you have steam. And what's driving it out through the exhaust is steam. It's coming out these little nozzles at more than a 1,000 meters per second, so the sound is like a high-pressure air hose or steam hose and puts out 110 decibels; it's a very sharp, piercing noise. When the fellow flies out here later today, it'll be surprising. It's not a roar or anything; it's a hiss—a very high-pitched sound."

My mind drifts away. Something Suitor has said sets me thinking, how the fuel at work here, hydrogen peroxide, is made of the stuff that is absolutely essential to human survival: H_2O_2. Water and air. This speaks to a paradox that has been wriggling around inside me but I've had a difficult time articulating. Now, it seems clear that flying alone and without wings is at once the most basic human urge, ingrained in our very essence, and it is also an incredibly difficult physics problem to solve. The Rocket Belt's fuel may contain the same molecular structure as water and air, but that doesn't mean launching ourselves is inherently within our ability. Indeed, getting airborne in this manner remains among the last great technological riddles we're still aching to unravel.

Journalist William Langewische once put it much better in his beautiful "meditation on flight," a book called *Inside the Sky*, pinpointing the paradox in this way: "Flight's greatest gift is to let us look around, and when we do we can find ourselves reflected within the sky. We find reflections of ourselves there, but of all the inhabited places the sky remains the strangest. Early evidence suggested perhaps it was meant to be so, that the sky constituted a sacred territory, and that if God had meant people to go there He would, for instance, have made them lighter than air." That helps explain why some four hundred years passed between da Vinci and the Wright brothers without a true breakthrough in mechanical flight, and why perhaps it is not too surprising that some fifty years have dissolved since Wendell Moore first sketched his sandy Rocket Belt, with minimal forward progress made since.

That, anyway, is what I got to thinking. My mind snaps back to the present just as the brunette, a producer with a Los Angeles PBS affiliate, asks Suitor, "Do you foresee a time when we'll all be like the Jetsons flying around?"

Bill is ready for this one. It may have been a few years since he was last in the spotlight in a significant way but, still, he's ready. "I hope not."

My stomach drops a bit, even as my nodding, grinning face says, "God bless the old straight shooter."

"And the reason I say that is: picture a man machine, maybe three hundred pounds flying through the air." I picture it. "I had several real close calls with wires that I just didn't see. And you're sitting in your backyard having a nice summer lunch, and some moron hits the telephone line and ends up on your picnic table, or they collide. During several dual flights, just doing that, I was always, 'Where's the other guy? Where's the other guy?'" He pantomimes looking over his shoulder. I hear what Suitor is saying, but it doesn't strike me as a good enough reason for why we aren't all using jetpacks right this minute.

The PBS reporter swoops in for her last question, "Do you have a favorite moment when you were airborne with these?"

Again, he's more than ready:

"I've thought about this many times. And maybe it's on that footage of me flying at Fort Niagara for the army. It was late in the day in June, and the angle of the sun was perfect for it. It was the first time I'd ever noticed it—I was flying over all those old buildings, and I noticed my shadow racing ahead of me. It was distracting. And it was the first time I'd ever gotten the feeling of what it is I was doing. Like—holy mackerel, that's my shadow! At twenty-one seconds, you don't want to pay attention to it too long. Then, I was away from flying for a number of years, and I was making a flight in Texas and it was my first check flight, again late in the day but in January. And I went around an apple tree, and as I made the turn I saw my shadow and I followed it and landed on my shadow. Those are a couple of the good ones."

He'd flown in front of more than a billion people around the world as part of the 1984 Olympic opening ceremonies. He stood in for Sean Connery, blasting off as James Bond in *Thunderball*. He's flown the Rocket Belt some twelve hundred times, every one of them twenty-one-second euphoria shots mainlined straight into his soul, the most seductive drug not on the market. Yet here is Bill Suitor reminiscing about seeing his shadow. His *shadow*. The simplest pleasure. My daughter Oona gets fired up about seeing her shadow and picks up a stick to watch it change.

Somewhere Amelia Earhart must be smiling. The winged woman who tamed the Fokker F7 once observed, "You haven't seen a tree until you've seen its shadow from the sky." Yes, it would be wonderful to ditch my Subaru for a jetpack and never look back. Yes, Bell's twenty-one-second flight time is an unfortunate scientific fact that once thwarted our marvelous climb toward the clouds. But in a life that will invariably be too short, maybe just a quick downward glance with the sun perfectly positioned over one's shoulder is enough. One flash of shadow. Maybe not, but it's a nice thought. And so God bless you, Mr. Suitor.

The PBS cameraman cuts off his light—the First International Rocketbelt Convention is about to begin.

I rejoin Joanna and Jofie, and we shuffle into an auditorium, which is half-filled with about 150 attendees, for the day's first presentations. The museum director welcomes us and introduces the panel assembled onstage, comprising former Bell Rocket Belt pilot John Spencer; engineer and company press liaison Bob Roach; the second Bell Rocket Belt pilot, Peter Kedzierski, who at the 1963 Paris Air Show snagged the enviable nickname "Bird Man"; and, of course, Bill Suitor, whom Jofie and I quickly take to calling "Bob Woodward," for his institutional memory and for his physical resemblance to the Watergate reporter.

Over the next two hours each man rises to the microphone and paints in broad, avuncular strokes his experience with the belt and the history of the development of the technology. They marvel at what they've seen, and I find myself marveling along with them, despite a certain lack of—of, what? Charisma? Public speaking experience? The guys drone on at times, and I can sense part of the crowd growing restless around me.

"It's a battle of L over D," Bob Roach is now saying. "Lift over drag."

I perk up. I've actually heard about this from my dad. It has to do with the amount of power a machine can generate versus the amount of weight it must pick up off the ground, while taking into consideration such things as wind or atmospheric pressures. Drags.

"And the human battle of lightness over darkness," Roach continues, and I could be wrong but I think he is about to get all metaphysical on us. "The Jet Belt won that battle."

It did? Now I'm lost.

But the pictures help. A series of slides and amazingly low-tech PowerPoint demonstrations flicker on a scrim at the back of the stage like a fading dream. There is Wendell Moore, circa 1957, bow tie, horn-rimmed glasses, buzz cut—the very picture of sepia-toned, future-looking science and that singularly American indomitable spirit of adventure. And there he is, later, testing out his machine in a hunter-orange suit, white crash helmet, and white knee-high boots, an outfit that would've had Ziggy Stardust's tailor taking notes.

A handful of edited-together early films and promotional reels begin. On a scratchy black-and-white print, Hal Graham, the Rocket Belt's Neil Armstrong, pilots the first flight free of a safety wire on the date emblazoned into every attendee's brain: April 20, 1961. Graham hovers about five feet off the rough grass at the periphery of the Niagara Airport. When the peroxide exhaust hits the cold morning air, two streaks of white vapor shoot from the nozzles. With a slight twist on his left handlebar, Graham angles the jetovators behind him and glides through the air as if in outer space. A new film shows another pilot gently slaloming between three golf flags. Then the clip of Graham flying ship to shore, the water rippling madly in his exhaust's wake, and saluting JFK at Fort Bragg in 1961. I feel a knot rising in my throat.

Some of the footage drifts so far into the realm of extreme kitsch as to be a bit absurd. In a montage taken from the mid-'60s featuring the flat color palette of the era, Suitor wears a tomato-red racing suit and straps into the belt as that familiar, measured, pack-of-Kents-a-day, the-future-is-here newsreel voice exclaims, "One of the most spectacular devices is the Rocket Belt—it has captured the imagination of everyone who has seen it. And thousands have been thrilled by the sight of a man floating through the air with no visible support."

A placard of blocky red letters on a white board fills the screen during a promo clip with the words "Life Saving Operations." For this daring maneuver, the pilot carries a life preserver to a swimmer who, it seems, is drowning no more than twelve feet from the banks of a still and shallow lake. The orange donut is dropped, whereby a heavyset guy, looking for all the world like he is on his way home from the track, standing onshore in a porkpie hat and sunglasses, tugs at the rope, dragging the swimmer to safety.

Next we have "Troop Support—Securing a Village," and then "Assault Mission." The clips show Suitor zipping over barracks and through the forest, and it is thrilling to watch. This is followed by "Mid-Air Hookup," "Liaison Mission," and "Laying Wire," all over a surprisingly funky, synth-driven, predisco soundtrack. Even so, it's beginning to dawn on me why the army let the contract expire.

Soon John Spencer is talking about reaction controls—those tiny rockets that Moore thought to affix to Bell's supersonic X jets to enable navigation outside of Earth's atmosphere.

Next, Moore's daughter Carolyn Baumet is at the mic, remembering, "My dad lived, ate, and slept the Rocket Belt." She points to a framed pencil drawing sitting in an easel by the lip of the stage. "In his dream, he thought of his design for the belt. One night he got up and sketched this rocket chair and woke up my mother and said, 'Hey, Mama, I need your signature to verify this.' Today, I'd like to donate this sketch to the museum." A standing ovation threatens to erupt.

It's nearing the lunch break and not a moment too soon, as I've missed most of the morning buffet. But first: what's this? From the back of the auditorium, up near the rafters, comes a spaceman in a shredded rubber suit and racing helmet, a jetpack, or perhaps a tuba made of foil, on his back. A ripple of expectation washes over the room. Who is it? What is he wearing, exactly? Is that—?

The helmet is painted gold, twin shooting stars streaking across each earflap. He's marching down the stairs with a purpose, the rubber pants slurping with each step. As he reaches the stage he turns, and I can see the name stenciled across the front of his helmet: Graham. Hal Graham!—known here this weekend by former colleagues and adoring enthusiasts as "His Eminence." So that wasn't a dead raccoon in his trash bag after all, it was his original Bell rubber suit, a firefighter's uniform, actually, and this homemade tinfoil jetpack.

Graham removes the mic from the stand and makes it clear right away that he's not like the others. Or anyone else. He starts most sentences with, "Okay, here's the gig . . ." and talks in salty, unscripted bursts. He rambles from topic to topic, occasionally fixing his stare on the crowd to make a serious point: "Wendell wanted us to know that the early demonstrations are scientific, not a circus." Of his original suit, now splayed open to high on his thigh from years of wear, Graham notes, "No liquids could get in, sure, but the bad part was that no gases could get out, either. So we had a rule: no burritos, boiled eggs, or beer up to three days before a flight."

Graham takes about ten minutes to hit the major highlights from a seventy-four-year life. The most captivating account is a story enhanced by props, a Rocketeer doll, and a scrappily constructed landing platform. With props in hand, Graham details the time he fell twenty-two feet from scaffolding during a Cape Canaveral demo and lay unconscious for half an hour. "I fell on my head," he summarizes, revealing the stitches in his helmet where it had cracked. "Now I'm retired," the world's most awesome grandpa suddenly concludes. "I work on lawnmowers and I love it!"

Then, just one more thing. Graham fishes a baritone ukulele from the back of the stage and shuffles toward the microphone. He introduces an "original composition," as if he's capable of any other kind. The room is as quiet as the moon. Mouths hang open in expectation. His voice is part William S. Burroughs, part Uncle Fester as he begins to sing:

> I am getting old and feeble now
> And I cannot work no more
> They put the old Rocket Belt away
> No more demonstrations in front of JFK
> No more flights at the U.S. Pentagon
>
> Oh, my Rocket Belt days are over
> My fame is fleeting fast
> The task before you people
> Is to improve upon the past

Wendell, Ed, and Ernie are up there looking down
Their spirits now are roaming with the blessed
Their efforts on the project
Brought them great renown
I tip my hat up to them
I wish they'd come back down

Oh, my fame is fleeting fast
Reporters never call
Haven't had a press release in years
But sometimes in the springtime
More often, in the fall
I remember certain details
After tossing back some beers. . . .

"I think I'm in love," Joanna whispers.

I look to my left and see Bill Suitor standing in the shadows of the auditorium. In his arms is Kathleen Lennon Clough, who's father, Tom, was Bell's staff photographer. Kathleen is weeping hard, convulsing, as Suitor comforts her.

It's definitely time for lunch.

Back out in the museum, I grab a wilting turkey sandwich and a packet of mayo, and mingle. I'm impressed by the press turnout. Besides PBS, the History Channel has sent a crew, and a freelancer is covering the weekend for *Slate*. There's also an international presence in the form of an impossibly tall, cue-ball bald, suavely dressed German journalist here for a big techie magazine called *PM* and a hip-looking English guy writing for Mazda's customer publication. I try to ID some of the attendees and am able to place a couple of faces. There's Nelson Tyler, tall and lanky with a silver halo of hair, the Hollywood camera-mount entrepreneur who built the belt that Suitor flew for the '84 Olympics, and there's Ky Michaelson, aka the Rocketman since 1951, the first amateur to fire a homemade rocket into space. In the spring of 2004, Ky and a small crew traveled from Bloomington, Minnesota, to Black Rock Desert in southwestern Nevada to fire a twenty-one-foot rocket seventy-seven miles into the sky. The marvelous device zoomed to 4,200 mph in ten seconds.

Ky's easy to spot in his black-and-red shirtsleeves number that's a cross between NASCAR pit crew and Friday Night Bowling League with his name stitched above the pocket. And in case you missed that, he also has his name emblazoned on a white baseball cap, fighting for space with a large patch of Old Glory and a bald eagle. Ky's face is that of a friendly ferret, replete with John Holmes–ian mustache.

He was born in 1938 in South Minneapolis and grew up next to an airport where World War II fighter pilots were trained. As a boy, he'd hot-wire model cars with tiny rockets built from home chemistry kits. When he was fifteen years old he took the money he'd saved from a paper route, lawn mowing, house painting, and other suburban entrepreneurial endeavors and bought his first car—a 1932 Ford coupe. Ky promptly replaced the car's engine with a Ford truck motor and three high-powered carburetors. He had just built his first dragster.

This led to a spot on Dick Keller's racing team. Keller and his crew set the land speed record on Utah's Bonneville Salt Flats on October 28, 1970. That day the Blue Flame, automobile as cruise missile, skittered across the sandy stretch of real estate at 630 mph, or about 60 mph faster than the cruising speed of a Boeing 747. The Blue Flame's rocket fuel was a combination of hydrogen peroxide and natural gases. Ky was hooked.

He's since made a career out of rockets. He spent a few years consulting and doing stunt work on *That's Incredible,* the pop-science television show from the '70s, and was Burt Reynolds's stunt double in the films *Hooper* and *Smokey and the Bandit.* Ky's enduring hobby has been to think of something, anything, and then attach a rocket to it: a chair, a sled, a bike, even a toilet. Ky claims the SS *Flusher* can hit a top speed of 200 mph, but he's yet to get behind, or really on top of, the wheel. For a Discovery Channel challenge, Ky and crew—his wife, Jodi, and a couple of friends—once hooked two M1419 rocket motors onto a bright-orange outhouse and blasted it into the sky. Because he could. He holds some seventy-two state, national, and international speed records, and you can find him in the *Guinness Book of Records* for the rocket-powered snowmobile he built in 1969.

Ky's latest creation is the "Robo Rocketman," a seven-foot, two hundred–pound remote-controlled robot with the head of the Rocketeer and the body of a stainless-steel linebacker of the future. On wheels. With a (nonworking) jetpack on its back. The Robo Rocketman has video cameras for eyes and an installed sound system, through which Ky can speak to students during demos.

The rocketing renegade comes by his unusual obsessions honestly. Ky's great-grandfather John Michaelson raised college tuition money in the summer of 1905 by building a seventy-five-foot-high, two hundred–foot-long wooden ski ramp in the middle of Minneapolis's Wonderland Park and then using it to hurl himself and his bicycle some fifty feet, over a picket fence and onto a smaller wooden platform. John advertised the feat with homemade posters that announced "See the Great Michaels in his all inspiring, death-defying leap across the gap." He'd shortened his name so his mom wouldn't know what he was up to.

Ky claims that relatives of his are responsible for creating the first motorcycle transmission and clutch, the rotary lawnmower blade, the flip-top aspirin box, the automatic remote-controlled garage-door opener, and oxygen masks that are still used on airplanes. He says his dad invented the Boroscope, a device that can peer inside the mechanics of an airplane's wings, giving technicians a glimpse of its innards. There is much I'd like to discuss with Ky, but for now I make a straight shot for the guys I want to meet first, the convention's most insanely committed attendees, the dudes who actually plan to build and fly one of these things.

Jeremy McGrane, a boyish thirty-two-year-old from tiny Raymond, New Hampshire, is so squeaky clean he says he's never had a single drink or cigarette. For the past ten years, since he caught an old video clip from the first Super Bowl's halftime show while working in an electronics store, his vice has been the rocket belt. "I was obsessed," he tells me. "I had to have one of those things—I thought that as soon as I saw it on TV."

He works during the warm months laying thick telecom cable deep in the ground on Nantucket Island, a job that pays him well but requires a seven-hour round-trip commute. To get to work, Jeremy wakes up at three and covers thirty-two miles of Atlantic Ocean by high-speed ferry. "I'd prefer nothing more than to strap on a jetpack and go," he tells me. During most winters, Nantucket terrain freezes solid, freeing Jeremy to retreat to his parents' garage and continue tinkering. He's done most of the building alone, teaching himself the machine's mechanics through sometimes agonizingly slow trial and error.

You can't argue with the results. His beautifully sleek, blue-corseted rocket belt stands out from among the other three amateur models on display. "I'm interested in the mechanical aspect, but to actually fly the thing scares me out of my mind. Now that it's almost done, it scares me. I mean, it's just steam holding you up." Nonetheless, Jeremy hopes one day to be able to parlay his efforts into a moneymaking exhibition operation akin to the one Kinnie Gibson has going with Powerhouse Productions, earning up to eighteen thousand dollars for half a minute's work.

For all the toil, he should see some rewards. His decadelong quest has led him down some peculiar roads. In 2002 he flew solo to see a guy named Avril Porter in Milton, Florida. Jeremy had stumbled upon the drag-car engine builder's name online and contacted him for motor-building expertise. "The things he does with metal are just incredible," Jeremy gushes. So he flew to Pensacola, rented a car, and began driving west. "The fields were getting bigger and bigger, and the campers were getting smaller and smaller. I was thinking, 'Oh, man, what am I doing?'"

Finally, he came to a turnoff on the road and pulled onto a plot of land housing a small camper. A big guy in a prison-orange T-shirt lumbered out of the camper and bellowed in a thick southern accent, "Glad you made it, hoss!"

Porter took Jeremy into a machine shop, and sitting on the ground was a rocket motor capable of sixteen thousand pounds of thrust. "That there's for a man to launch himself into outer space," Porter said.

Jeremy stammered, "Really? What's going to happen?"

"He's gonna get keelt."

Despite, or maybe because of, this auspicious introduction, a deal was struck, and two months later a decidedly much smaller engine arrived in the mail. Jeremy's project was taking off. Eventually, he befriended a NASA optical engineer and a fellow rocket-belt enthusiast named Mark Wells (the same Mark Wells who had known Thomas Moore). Together, they slowly re-created Wendell Moore's machine, improving on materials and construction where they could.

By the time the rocket-belt convention rolled around, Jeremy was gearing up to fly and had already purchased thirty-six gallons of fuel for twelve thousand dollars from Erik Bengtsson, a Swedish chemical engineer out of an Ingmar Bergman casting call who's one of the few people in the world capable of shipping rocket-grade hydrogen peroxide anywhere in the world. As I speak with him under the florescent bulbs of the Niagara Aerospace Museum, Jeremy tells me he is convinced he'll be in the air before the end of 2007. (That didn't quite happen, but he hasn't given up yet.) "Most guys are dreaming about alcohol and women—not me. I'm just dreaming about how to make a throttle valve. It's peculiar behavior, I'll admit it, but sometimes I can't sleep at night."

The toll this has taken is not lost on Jeremy. "I could have been a surgeon by now, if you consider all the time I've spent on this. This takes away from being with my family, from being with my girlfriend.

I don't ever get to just go out and enjoy the day. The newness is wearing off a bit—working on the weekend is feeling more like work."

Still, my guess is he'd say it's worth it, if for no other reason than the fact that, from scratch, Jeremy McGrane once constructed a throttle valve almost identical to the one used by Bell Aerospace. "It was within one–one thousandth of an inch of what Bell had. So that's pretty cool." Over the course of my search, I'll hear tales similar to Jeremy's repeated, with a new wrinkle or two thrown in here and there, by several other garage tinkerers all over the world.

Take fifty-three-year-old Gerard Martowlis, standing in front of me now, a Boba Fett T-shirt straining against his thick middle, clutching a stack of homemade bumper stickers that read, "I'd Rather Be Flying a Rocketbelt."

The chemical-waste expert has been smitten with jetpacks from very early in life. As a ten-year-old boy, he rarely missed an episode of Commando Cody's black-and-white, jetpack-fueled adventures. Growing up, Gerard was a hard-core space geek, constructing models of the Gemini until way past bedtime. But it wasn't until he came across a *Popular Mechanics* article on rocket belts in the late '80s that Gerard realized he could, possibly, make his dream come true. He began doing research at the local library. Not long afterward, he went to the Smithsonian's Air and Space Museum, chaperoning one of his daughter's class field trips. He looked forward to seeing the Wright brothers' plane and an X-15 jet; he didn't know at the time that the museum also exhibited a Bell belt.

Gerard has lived in the same house in Rahway, New Jersey, since 1981. He divorced twenty years ago, when his daughters were only four and five. "I'd tuck the girls into bed and then sneak down to the basement to work on the belt. It helps to have something else to dwell on."

For two decades Gerard kept his project under wraps—only his brother, a brilliant welder who assisted with some of the construction, really knew what he was up to. When the gas company's meter reader came around, he'd hang a sheet to hide the machine.

His work as an industrial-waste technician gives him a leg up in understanding the belt's fuel structure, but he's had to find help with some of the engineering issues and machine work. His stock answer if anyone wonders why he's prying is "I'm building a pressure washer for blasting gum off the sidewalk or cleaning the side of the house."

Gerard is now engaged to a small, pretty woman named Serafina Giordano, who accompanied him to the convention. As he and I speak, she stands nearby, rolling her eyes affectionately as her betrothed details his mania.

A quick digression. In the few months following the convention, a newly energized Gerard went on a diet that consisted of a lot of plain tuna fish. He dropped 15 pounds, down to 176, in anticipation of soon flying his device. Then, on New Year's Eve 2006, at three in the afternoon, Gerard Martowlis went into his backyard, strapped his machine to a sturdy locust tree, and, for the first time, fired up the motor. Later that day, he reported the results to the Yahoo! group: "Even though the test was run at 43deg.F and a low fuel tank pressure of 300 psi, the thrust it generated had the belt straining on its restraints. . . . I must admit the butterflies were doing a dance in my stomach when I twisted that throttle earlier today. I'm still pumped from the experience, WOW these things have such tremendous power."

I'd heard from some Yahoo! group members that Gerard's throttle valve looked different from others they had seen, so I asked him about it. He e-mailed: "My Throttle valve is an off the shelf S/S ball valve that's been modified internally and externally to allow it to be finely adjustable. Those modifications provide a positive feeling at the twist grip controls as well as providing a positive mechanical method of closing the valve."

Back in Niagara Falls, I am soon introducing myself to a Floridian named Tom Edelstein, standing near the half-ring of rocket belts. Tom looks like a lion tamer, or maybe just a lion—all stocky build, long mane of blond hair, thin lips, and tightly set jaw. Under his arm he holds a thick binder full of papers. The pages tell, in painstaking detail, the sad story of how Tom had made great progress a few years ago with his own rocket project, even managing a few short, tethered hops off the ground, only to see his dream literally go up in flames when a batch of contaminated fuel started a fire in his workshop. His rocket belt took the worst of it. Tom had documented every last test, detailing how each part had performed in notes that would've made Wendell Moore proud: the weather conditions, how well the throttle fired, how the fuel responded to its catalyst. He still doesn't have the heart to clean out the workshop. The charred flying machine remains in the shadows. Tom's throat catches, as he tells me, sotto voce, "My wife doesn't know about this, but I have every intention of rebuilding it." His wife—tall with former-cheerleader good looks, wearing mom jeans—was inspecting the rocket chair nearby.

If anyone should become a real-life rocketeer, it's Tom Edelstein. He joined the family business when he followed his parents into the Ringling Brothers circus at age five as a flying trapeze artist. At twenty he set a Guinness Record for the most midair flips. "It's called the triple twisting double—it's five revolutions," he says. He met his wife, Linda, on the job; she was his trapeze partner. Now Tom works as a stunt de-signer for live shows, "Cirque du Soleil–type performances, stunt ef-fects where people fly through the air and so on and so forth."

I break away to collect my thoughts. While Jeremy, Gerard, and Tom all told fantastic, heartwarming, inspiring, and also occasionally poignant and sad stories about chasing dreams, overcoming obstacles, and throw-ing caution very much to the wind, something, I couldn't help thinking, is missing. That something, I begin to realize, is immortalized by the convention's slogan, stitched into khaki-and-blue baseball caps, between a pair of cartoon rocket men: "Where the Past Meets the Present." Meets the what? Where the past meets the . . . present? You sure about that? What about, you know, the future? I mean, is this all one big nostalgia fest? Have we all just thrown in the fuel line and admitted that we'll never be able to come up with something that flies longer than a television spot for adult diapers? And by we, of course, I mean them. Where the past meets the present? Sorry, but that's just not going to cut it. We need to bring the future into this discussion—and fast.

In fairness, Bill Suitor had alluded to it in his morning presentation when he briefly stopped his history lesson and proclaimed at the podium, "We've got to get away from rockets—and get into jets!"

Yes! Jets. Jetpacks. And that's another thing—this whole rocket-belt problem of semantics. "We're not doing ourselves any favors on the PR front, people" is what I want to tell attendees. I know, I know—that's what Wendell Moore called the machine, and so that's what it is. But the public has spoken, is speaking, and what they want is *jetpacks*. Not rocket belts. So let's lie. Let's call the thing a jetpack, and I guarantee at the next convention you'll have way more than 150 retro-minded boomers talking egg timers to track flight times. Where's the harm?

When I'd once mentioned the rocket-belt name situation to my advertising friend Greg, he scrunched up his nose and said, *"Rocket belt,"* as if the very words stank. "Well, that sounds a bit swishy, doesn't it?"

Yes, yes it does. This is not to belittle what you've done, Jeremy, Gerard, and Tom, but what might happen if you poured all that blood, sweat, and dough into an actual *jetpack*?

Just as soon as Suitor had raised hopes this morning with his mention of jets, he shot the idea down. "So, anybody know Bill Gates? Because it takes so much darn money."

Argh!

But then, the whiff of hope. Or magic, maybe. I drift back into the main convention room near the auditorium. Joanna is chatting with a guy who definitely stood out from the pack. For starters, he was about half the age of the next youngest person in the room. He wears a student's scruffy beard, a jaunty hat that Indiana Jones might've packed while vacationing in Australia, and a fitted, stylish army-green jacket.

In a room ruled by saggy jeans and ill-fitting tees, the kid stands out; he has a look.

Joanna excitedly waves me over. "This is Will Breaden-Madden, he flew over by himself from Ireland—you have to hear what he's working on."

We shake hands, Will stepping back slightly and lowering his head as if doing an interpretive dance about shyness.

Of course I want to hear all about what he is working on—the little bit I know sounds very promising, but I wondered how much to believe. Will and I had exchanged a few e-mails soon after I'd joined the Yahoo! group. He told me he was studying theoretical physics at Queens College in Belfast. He told me that when he received his pilot's license at a flying school in Tampa, Florida, he became, at seventeen, the youngest Irishman ever to do so. And he told me he was working on the flight-duration problem that had plagued Wendell Moore and that he hoped to extend flight times through the use of jet engines. He told me he was building a machine and he'd given it a splashy, if unintentionally funny, name: the ShamRocket.

But that's all he'd tell me for now. Unlike most of the group's members, Will prefers to keep a low profile, to fly under the radar, as it were. At least until he has something to show.

"I'm planning a trip to the UK to see Stuart Ross later this fall," I tell Will now. "Perhaps I could swing by to see you, too."

Stuart Ross is a commercial airline pilot, living on a one-acre farm in Sussex, England, an hour train ride south of London. He'd already begun testing his rocket belt on a safety tether, and so he is something of a celebrity at the convention. *New Scientist* magazine once profiled Ross. I want to see his shop—and to find out if he might be working on anything else.

"Yeah, yeah, absolutely, that would be fantastic," Will says, full brogue flying. "Though I should warn you that I don't expect to be finished with the ShamRocket by then."

"That's okay." I play it cool. "I'd just love to hang out a bit and see how things are going." I didn't tell him that we also needed to have a word about another unfortunate name—ShamRocket. Talk about a PR problem. We promise to stay in touch as my trip nears.

The afternoon's presentations are cut short, and I'm not sure anyone minds. Threatening storm weather is gathering outside, so the rocket belt demo has been bumped up to midafternoon. This, after all, is the moment—okay, the half a moment—everyone is waiting for.

Eric Scott is a lean, craggy-faced former stuntman and air force pilot who once trained at the nozzles of Kinnie Gibson. When Troy Widgery, an energy-drink entrepreneur from Denver, needed someone to pilot his Go Fast! promotional belt, he poached Scott from Gibson. As the convention is unfolding, Gibson is allegedly in the process of suing Scott for breach of contract to the tune of $750,000.

As far as I can tell, Go Fast! is the convention's one and only sponsor. At three in the afternoon, Scott stands on one end of the closed-off street in front of the museum. He wears a black-and-red racing suit and a black motorcycle helmet with a clear face shield. The sport drink's swooshy logo is splashed across his chest, the rocket belt's corset, and down the side of its bloodred fuel tanks.

I find a place along the sidewalk and am surprised to see about five hundred other spectators gathering on either side of me. Grandparents with grandchildren. Skate kids. Mothers of three. Where have they all come from?

A bushy-haired, wide-bodied young man whose T-shirt announces, "I write code so you don't have to," takes a spot behind me. I ask him what's brought him here. "I saw these fly in the *Rocketeer,* but I've never seen one fly in real life." Then he catches how earnest that sounds. "I've got a couple of them and just want to make sure he can fly as good as I can."

I'm hoping to record Eric's flight on video. Despite several warnings about the 110-decibal piercing shriek of the steam exhaust—like the world's biggest balloon losing air quickly—I've failed to bring earplugs.

This proves to be a problem. Eric gives a thumbs-up. I am about to see a man. Fly. Through the air. I peer at the viewfinder. The pilot cracks the throttle open, and my eardrums melt. I grimace and clutch at my left ear. It's softened candle wax. With a rush of wind flapping his pant leg, Eric Scott lifts off the ground as if by the Hand of God. I struggle against the noise to keep him framed in the shot. Suddenly, he is very high in the air—thirty, forty feet above us? Higher?

He zips over the museum's broad awning, over the Huey Cobra helicopter standing guard above the entrance. It doesn't seem real. As with most things that are too wonderful or too dreadful to fully compute as they are happening, I am not really sure I am seeing what I am seeing. It's as if I've become an instant conspiracy theorist—is that a wire I see holding him up? But no, there weren't any wires, just superheated hydrogen peroxide burning on a man's back at 24.5 ounces per second, forced out of stainless-steel tubes fast enough to create four hundred pounds of thrust. That's all.

Eric reaches the end of the block and descends the air. When his toes dangle six feet from the pavement, he hovers for two beats and then drops softly back to Earth. The crowd, breath holders all, exhales, erupts, and rushes the pilot as if we've just won the seventh game of the World Series, as if Eric Scott were Larry Sherry. He'd been off the ground for fifteen seconds.

The PBS crew jockeys with local news cameras for the best angle on Eric. A six-piece band—when did *they* set up?—breaks into "Dancing in the Street," though no one is.

"It's a dream that's reality," Eric is telling PBS. "It's so hard to explain what that's like. It goes against everything you've known because you're defying gravity."

Steve Agnew can't get enough. He's beaming like a parolee over my right shoulder as Eric patiently takes all comers. Steve is forty-two years old, but with his wide eyes, flattop, and backpack, he looks like the oldest corn-fed seventh grader ever. He punctuates Eric's sentences with a "Geez!" here, a "Wow!" there. Steve tells me he saw Suitor on television, flying at the '84 Olympics. "Before that I thought they weren't real." He was with his wife on the Hawaiian island of Kauai celebrating their nineteenth wedding anniversary when he'd gotten wind of the First International Rocketbelt Convention. The couple hopped on the next plane, picked up their three sons at home in Yakima, Washington, and headed straight for Niagara Falls. They traveled for thirty-six consecutive hours to make it.

But that's a small price to pay for a guy who earned enough working for a "fruit sanitizing" chemical company to pour about $100,000 into his own rocket-belt project and who now tells me that when he found a computerized plating machine that usually sells for $150,000 for $5,000 online, he flew across the country to collect it and drive it home. He spent three days parked in a Home Depot lot, building a wood crate large enough to contain the device in order to transport it. "That was a blessing," he tells me of the deal.

Steve's wife, grinning behind wide sunglasses, sways to the beat beside him. She doesn't appear to mind the change in plans. "Whatever makes him happy," she says, massaging his shoulder. She doesn't need to finish the sentence.

But I have to ask Steve one thing: "Why?"

He smiles, his mouth packed with big, healthy Washington State teeth. "Just because it's something you can do for yourself." I want to remind him there are plenty of cheaper things you can do for yourself, but then he adds, "Oh, and for the bragging rights."

Day 1 of the convention is winding down. I excuse myself and find Joanna and Jofie, both pumped from the demo. I'm exhausted, but my brain's buzzing and I've barely eaten all day. Jofie suggests a genius plan for the evening. We'll round up a few of the most promising characters we can, take them to dinner, get them drunk, and find out which one is really working on a secret jetpack.

About an hour later, in the middle of another Niagara city block dedicated to low-cost dining, dark bars, and Laundromats, we pull up in front of La Hacienda, a family-style Italian restaurant. Our party consists of Ky Michaelson and his friend of thirty years named Jeff, who wears a perpetually bewildered Warholian expression and a T-shirt of the cosmos that says, simply, "Space Time"; Bill Higgins, a fifty-two year old with an impressive cascade of straight brown hair worn in a precise bowl cut, who operates a Web site for flying-car enthusiasts; Josep, a city water department employee from Barcelona, Spain, who has left his wife and son behind for three days to follow a dream he once found in a library book; and, naturally, Will Breaden-Madden. We commandeer a long wooden table in the front, under faux stained-glass-shaded saloon lamps dangling from the ceiling.

Jofie's genius plan hits an immediate snag. No one besides Josep and us is drinking. Will tries to order a banana daiquiri because he is nineteen and away from home, but La Hacienda can't do a daiquiri. We order a bottle of Chianti for our end of the table, and before it arrives I've learned the following about Ky Michaelson:

1. He throws an annual Super Bowl party at his home in Minneapolis.
2. His home in Minneapolis has a swimming pool in the living room.
3. He serves eighty pounds of shrimp and eighty pounds of lobster at his annual Super Bowl party.
4. He's inviting us to the next Super Bowl party.
5. As the first amateur to put a rocket into space, he beat the winners of the X Prize in 2004, the first amateurs to put a rocket ship into space, by twenty days.

6. He knows Prince and Bob Dylan.
7. He once dated Morgan Fairchild.
8. He "used to be good-looking."
9. His Web site, www.the-rocketman.com, contains "over five thousand photos and two thousand videos."
10. "If you are ever in town and want to see a four-mile particle accelerator," call Ky.
11. One of his three kids is a boy named Buddy Rocketman Michaelson.
12. Yes, that's his legal name.

I want to pick Will's brain about his ShamRocket but am too distracted by Ky, sitting directly across from me, crumbling crackers into his minestrone and then slurping it down with both hands on the bowl, needling Jeff for clogging the restaurant's men's room toilet. Gabbing loudly on his cell phone and then shoving it in my direction, he asks, "Hey, want to talk to a rocket woman?" At which point I find myself talking with Mrs. Michaelson, just back from a trip with the kids to Yosemite where everyone had a great time. By the end of the meal, only one thing has been decided for sure: in the winter, Joanna, Jofie, and I shall travel to Minneapolis to watch Super Bowl XLI at Ky's place.

Back at Peter's that night, my mind is humming with a million questions. Is Will really building a jetpack? Is he the only one? Is Bill Suitor right that these things are too dangerous for the average dude? Will the real future ever arrive?

I find a quiet patch of rug in the living room, lie down looking at the ceiling, and dial home. Catherine answers. She sounds tired but still giggles as she updates me on Oona and Daph. "Daphne's getting so big!"

"I've only been gone a day, babe."

"I know, but I'm pretty sure she grew a lot last night."

I feel out of sorts. Everyone's always saying you shouldn't miss out on your children's first few years. It passes in a flash and can never be brought back. Appreciate every minute. And here I am, clear across the state, breaking bread with a table full of madmen. Part of me just wants to find my damn jetpack already and get home. But it is becoming obvious that I'm not going to bring home the mother of all Niagara Falls souvenirs—not this weekend, anyway. My quest is just beginning.

We say goodnight and hang up. I'm exhausted—this dream-chasing business will wear you out. I limp across the living room, where I can see through a sliding glass door the shadowy outline of Peter's very handsome and very large barbecue on the redbrick backyard patio. It's so quiet out there. The life Peter and Diana have built makes a compelling case for suburban domesticity. At that moment, I badly want to bring Catherine and the girls out to Buffalo to set up a sweet, simple life. We could get teaching jobs. Away from the hypercompetitive Manhattan media world, we could finally let ourselves go and wear sweats most of the time. I could buy an easel and take painting classes. Or maybe throw clay. Oona could learn to skateboard. Daphne could, well, she could keep growing big, and then learn to skateboard, too. It isn't jetpacks—but it sure sounds pretty good. I squeak my way onto an inflatable air mattress and pass out cold in about twenty-one seconds.

The next morning, before the lectures begin, I wander into the museum's conference room. A long, lacquered oak table stretches nearly the length of the twenty-foot room. On one end of the table a pile of commemorative hats mocks me with the slogan "Rocketbelt Convention 2006: Where the Past Meets the Present."

At the far end of the table a round fellow in a black turtleneck manipulates two joysticks to control a tiny jetpack on a laptop screen. The speck of light flits all over, soaring then diving above a pixelated metropolis. Bill Higgins sits, watching the action. "I'm not used to this throttle control at all," says the guy at the controls.

Bill squeezes his face into a half-squint, half-smile and says in a deliberate and modulated tone: "Well, no one really is, except for a few people at this convention."

Day 2 is a lot like day one—morning lectures and slides, lunch break, more talks, and then an afternoon Go Fast! demo. A tease really, but a satisfying one.

Some highlights from inside the auditorium: Science writer Barry DiGregorio on Wendell's career. Black-and-white memories of the dawn of the space age. Wendell's work on the supersonic, needle-nosed X jets with a dashing young pilot named Chuck Yeager. Crew cuts, cigarettes, sharply lit conference rooms seemingly borrowed from the *Dr. Strangelove* set. DiGregorio then plays a clip from an old radio program. It's Moore himself, his voice a gravelly staccato, risen from the grave, courtesy of the low-tech recording device DiGregorio holds to the microphone. "You feel as though you are being lifted—as if by something that held you under your armpits." Okay, it isn't exactly Homer, but check this out: "It very certainly has a place in the world for helping humanity."

Yes!

"Eventually I believe these will be used as commonly as a second car—to go to work or to the store." And there it is. From the inventor himself, clear evidence that the future is overdue.

Later, aerospace engineer Doug Malewicki blitzkriegs through a bizarre and highly entertaining presentation about, well, I'm not sure what exactly. But there are slides of the board game he created in 1965 about nuclear war called . . . Nuclear War, blush-worthy photos of his daughter in quasi-medieval garb (that is, tattered khaki miniskirt and matching tube top) as the fantasy warrior Mischa Redblur for a series of comic books, and, most spectacularly surreal, shots of his Robosaurus. If you are male and love fire, robots, and fire-breathing robots capable of crushing entire real automobiles in their stainless-steel teeth, then you may already know about Robosaurus. In 1990, Doug thought it would be a good idea to build a $2.2 million, 40-foot-high, 58,000-pound "electrohydromechanical monster robot" spitting 10 feet of real flames. The world seems to agree. Robosaurus has been performing at car shows and other testosterone-dipped events across the globe ever since, and was even satirized on an episode of *The Simpsons*.

Then it's back to rocket belts and Stuart Ross's tender film chronicling his decadelong quest to get airborne. First, a montage of Ross around the yard, mowing, raking, and bagging leaves. Soon he is firing up his throttle valve. On the soundtrack, Roger Daltry is vowing that he "won't get fooled again." Pete Townsend's guitar crescendos, and, with the drama sufficiently heightened, we see a shot of Ross's toes, encased in white Nikes, dangling maybe fourteen inches off the ground. This is an image that will plague me and I think represents, in many ways, the sad, noble efforts of these would-be rocket men. Ross has pumped so much of his own cash into this project he'd rather not keep a tally. He's acquired a welding machine the size of an oven. He's spent countless hours tinkering in solitude. And what has it gotten him?

Oh, no, that's not all—it has also nearly gotten him killed. On the screen now we see the determined pilot in flight suit and helmet, belt strapped tight. (It dawns on me that half the reason certain fellows do this is because it gives them an excuse to wear that sweet pilot's suit and brain bucket.) Ross walks the long wooden platform he's built out back, a safety wire clipped behind his head. His crew, a single mate, looks on anxiously. He should be worried. This time, when the throttle is cracked, it unleashes a nightmare. The homemade part, so agonizingly crafted over days, weeks, months, breaks. The valve sticks open, a hydraulic lock, allowing too much fuel to rush into the combustion chamber. Ross loses control completely. He is whipped about like a plastic bag in the wind. He smacks into a nearby wood fence—and bounces off. His crew member tries to grab him but instead gets a mouthful of Ross's helmet and is knocked back through the air some fifteen feet. It feels like the longest twenty seconds on film. Finally, the fuel runs dry; the clip ends. Miraculously, no one is dead. In fact, the worst of it is Ross's badly bruised knee. His friend's chronic back troubles, the Englishman swears, were fixed on impact. The lights come back on. Ross turns to the riveted room and dryly notes, "The key to the rocket belt is the throttle valve."

I need some air. My crew and I walk over to the falls. Jofie, ever the optimist, wonders if that day's Go Fast! exhibition will up the ante and take place there. Staring at the mad rush of water, I certainly hope not. As impressive as it was to see a man soar yesterday, I dread the thought of what might happen if Eric Scott attempted an Evel Knievel on the 1,060-foot-wide wonder, especially after that harrowing Stuart Ross video.

I will always associate this spot with Christopher Reeve's *Superman II*. I was nine years old in 1980, the year it was released. The flick gave me my first glimpse of Niagara Falls as our hero zipped down its cascading facade to rescue a young boy who'd gone over the rail. Despite Superman's religion-revealing tights and the not-very-menacing red cape, it seems unspeakably cool to be able, whenever the impulse struck, to just fly like that. Up, up, and away. Unencumbered. Uncomplicated.

Now, a few months into my jetpack quest, the future seems anything but uncomplicated. It seems our best shot came nearly fifty years ago, and who knows what lies ahead? Standing next to the thundering falls, all I can think is that from what I've seen so far, there is very little reason to believe you, me, or anyone else is ever going to get their rightful jetpacks. How would I ever to break it to that Craigslist guy?

The second Go Fast! demonstration is about to begin. Again, I find myself standing near Steve Agnew, his immense enthusiasm a human
magnet. "That thing's just pretty," Steve gapes, as Eric preps for takeoff. "Look at the detail in the corset—carved fiberglass!"

Again, Eric soars. The same glorious arc, the same stunning unre-
ality of it. The same World Series surge by the crowd to greet him on
the far end of the block. This time, though, I'm ready with balled-up
tissue paper for my ears and so am better prepared to capture it all
on video.

By the time Eric's feet touch back down on concrete, I can already
feel the weekend slipping away. Or maybe it's the homeward tug causing a gravitational pull away and out of Niagara. But first one more
spin upstairs, one more look at the belts. For all I know, this might be
as close as I ever come to my jetpack.

I enter the museum's back room and find Tom Edelstein and Gerard Martowlis comparing notes. On a TV screen in front of them, Tom's
tethered attempts play on a loop from a past with no future. All five
foot, three inches of him, sleekly outfitted in black racing suit and
matching helmet, straining against cruel physics. His lion-tamer boots
dangle barely above our intensely magnetic Earth. "I think I'm the first
American amateur to get off the ground," Tom says, fixated on his televised image.

Fifteen seconds of silence.

"What's the capacity of your tanks," Gerard asks softly, "if you don't
mind my asking?"

"Less than five."

"Less than?"

I struggle with the math. Less than five gallons—if the fuel is burning at, let's say, twenty-five ounces per second, that gives Tom, in the
best-case scenario, twenty-six seconds in the air—if he's lucky. And so
far, at least where the rocket belt is concerned, he isn't.

We make great time getting back to Brooklyn. I'm speeding, racing to get home. Even though Oona and Daphne will have gone to sleep long before I get back, I cannot wait to kiss the tops of their heads, to inhale the sweat-sweetened smell that is pure parental intoxicant.

My crew has increased by one. Over the course of the weekend, the *Slate* writer, an enterprising, pompadour-rocking badass named Larry Smith, and I had chatted a bit. It turns out he lives not far from me in Brooklyn, so he's hitching a ride. It's a good pickup, as Larry proceeds to entertain us for most of the eight hours back with stories about his wife's time in a federal prison (youth, indiscretions, Indonesia, drug smuggling, etc.) and with accounts of his own time spent on the jet-pack trail.

Earlier in the year, Larry scored a *Popular Science* assignment and traveled to Cuernavaca, an hour-and-a-half drive south of Mexico City, to interview rocket-belt legend Juan Lozano, who apparently is assembling a small arsenal of flying machines and other rocket vehicles. I'd read about Juan's exploits—he's even flown on a tether a few times, and recently he sent his daughter up as the first woman to test a belt—and was in the process of making travel arrangements to see him myself.

"Oh, you've got to stay with Juan!" Larry practically shouts. "If Juan offers to put you up—do it. You won't be sorry. They've got avocado trees, orange trees. You'll be sipping tequila in the pool. And good tequila, too."

This is excellent news. If one thing is clear to me in the few hours after the First International Rocketbelt Convention, it is that I could really use some help. I am, at least scientifically speaking, in way over my head. I can't change the oil in my own car, and I want to fly a jetpack?

I need a mentor, someone with an excellent working knowledge of the important principles of physics. An expansive understanding about the history of flight wouldn't hurt. Neither would thoughts on thrust, torque, yaw, pitch, roll, center of gravity, and, of course, throttle valves. And if he is available to take my calls at all hours of the day and night, to help me get my mind around certain principles, or even just rhapsodize about flight, well, all the better.

But who am I kidding? Where on Earth would I possibly be able to find such a person? I stomp on the gas, the thundering falls receding into the dream of my past.

CHAPTER
4

Out of This World

As we go up, we go down.

—Guided by Voices,
"As We Go Up, We Go Down"

The week after my trip to Niagara Falls, hundreds of thousands of gadget-ogling tech heads pack the 760,000-square-foot Javits Convention Center on Manhattan's far West Side for *Wired* magazine's NextFest. Begun in 2004, NextFest promises a glimpse at the magical and titillating technological near future.

The '06 edition includes ballroom-dancing robots, which Tohuku University, near Tokyo, spent about $300,000 developing; a stuffed seal named Paro, whose full-body sensors allow him to giggle if you tickle his whiskers or look sad if you smash his face; and something called the Hug Shirt, which uses Bluetooth wireless technology to let long-distance couples grope one another.

Of all of the 130 demonstrations, though, the runaway hit is Sir Richard Branson's SpaceShipTwo, a six-passenger, two-pilot suborbital minishuttle. Branson, the self-proclaimed "rebel billionaire," is the head of the monolithic company Virgin Atlantic. An avid extreme adventurer, parachuter, and closely cropped goatee wearer, he is one of a new breed of insanely rich moguls who've got a serious jones for space travel and are doing something about it. Microsoft's Paul Allen, Amazon's Jeff Bezos, and Branson are at the forefront of what is, in 2006, the nascent space-tourism industry.

To build SpaceShipTwo (and thus demonstrate his facility with counting), Branson hired Bruce Rutan, the legendary aerospace designer behind the $10 million X Prize–winning SpaceShipOne, the first privately built craft to take a civilian into space. Branson's space machine is three times as big as Rutan's first effort and has been described as "fucking awesome" by at least one tech Web site. For a mere $200,000, you can buy a ticket on SpaceShipTwo for the two-and-a-half-hour, eight-mile-high, 7G ride of your life. Despite the steep cost, judging by NextFest at least, it doesn't appear that Branson will have trouble filling the seats whenever missions begin. His vessel, designed in a way that resembles a slender pelican with rockets for legs, is the buzz king of the fair, and there are reports of average fellows trading two million Virgin Atlantic frequent flyer miles for a single seat on the ship.

Which is all terribly exciting, unless you'd rather fly a jetpack. But there is nary a 'pack to be seen at NextFest, where the future, it could be said, meets the present. This reinforces a notion that I can't seem to shake, which is that it all comes down to 1969. Ever since that fateful year (and I'm not even counting the Mets-Orioles World Series here), it's been nothing but space this, space that, space, space, space. That's no accident. Let's recap. On May 26, 1969, Bell releases to the public for the first time images of its in-development Jet Belt. But the military is already cooling to the idea and casting an anxious eye on Vietnam. Three days later, the godfather of the jetpack, its biggest advocate, Wendell Moore, suddenly drops dead of a heart attack. And less than two months after that, Neil Armstrong is doing laps in the Sea of Tranquility. A year later, Michigan's Williams International buys up the Jet Belt's intellectual property and prototype and decides

CVS/pharmacy

DAVID FREDRICKSON

Save $4 off your Purchase of
$20 or More
(Up to $4.00 value)

Expires 06/26/2010

4686662O1

1607/09/5620

there's more potential in Tomahawk missiles. The future is no longer on Earth; it's in space.

And there it has remained, lo these many years later. Still, I have to believe, guys like Branson and Rutan would, at the very least, have a soft spot for our unrealized jetpacking future. They, too, were once just boys with dreams, right? And together they had done some wonderful things with fuel. In 2005 their GlobalFlyer airplane circumnavigated the world without once stopping for gas. If anyone knew how to extend jetpack flight times and build a product that perhaps one day could be made available to schmos like me, it was those two guys. I make a mental note to contact Branson.

I have to put all those thoughts on hold, though, because life is calling. John Lennon once famously said that life is what happens when you're busy making other plans. I am beginning to think, however, that life is what happens when you're not hunting for a jetpack.

Life is now taking me, Catherine, and the girls out west, to California, for a two-wedding, three-city, sixteen-day, one-bedroom exercise in extreme socializing and sleeplessness. We are first flying to the Bay Area where in about a week one of Catherine's nine brothers (Jimmy, the barrel-chested, good-natured toy-business entrepreneur) is getting married in the hills of Albany, near Berkeley. We'll split time between my dad and stepmom's place in Walnut Creek, a quickly yuppifying midsize town twenty minutes north of Berkeley, and with another Crawford brother (Joey, big smile, big heart, father of three girls) and his family in San Francisco. A week later, Catherine's old friend Nancy the Very Awesome is marrying the charming and hilarious Rob in Los Angeles. And we're going to do it all on a budget. Anyone who has kids knows this will be a pretty complicated maneuver to pull off. Even those without children can probably appreciate what is at stake here: the flying, the car rentals, car seats, diaper bags, sippee cups, bottles, diapers, sunblock, strollers, sleeping arrangements (in one room, two beds and a portacrib), nap schedules, rehearsal dinners, awkward chitchat, and gridlock.

Which is why I think this would be the perfect time to also check out some other flying machines in the area. I mean, a guy's got to cover his bases. At this point I have good reason to wonder if I'll ever really see a jetpack, let alone fly one. And bringing one home to Brooklyn to use the next time there's a transit strike, blackout, or terrorist attack? It was time, I'm sad to report, to start thinking about Plan B.

There are a handful of places in this vast and ambitious day-dreaming country where people are still behaving like it's 1952. I'm not talking about hoopskirts and fedoras; I'm talking about flying cars, flying bikes, and other wondrous machines meant for the air.

One of them is a zesty Nevada car-stereo magnate named Woody Norris who started a company in 2000 to work on what he calls the AirScooter. Part motorcycle, part helicopter, the AirScooter, while potentially able to climb as high as ten thousand feet, is meant to be flown closer to four hundred feet in the air. It is capable of flying for an hour and going the FAA limit of 55 mph.

Norris and his small crew used "coaxial counter-rotating dual rotor systems and space-age material technology" to construct the machine. In the spring of 2005 Norris said he was close to offering the AirScooter for sale to anyone with $50,000 to throw around and a sterling set of enormous balls. But by the time we are planning our trip west, Norris's device still doesn't appear close to market.

Meanwhile, down in Wichita Falls, Texas, a former Bell engineer named Jay Carter has put together a prototype of his CarterCopter, which is like a small piston-powered jet with the back lopped off, replaced by another propeller. It's designed to take off in the same manner as a helicopter—straight up—but fly like an airplane at about 230 mph. In the future, Carter plans to build a version that has foldable wings and can take to the streets, too. That's right—a flying car. "It's the holy grail," Carter once told *60 Minutes'* Bob Simon. "You know, that has been the desire since we started flying, to be able to take off and land vertically and then fly fast. It's going to change transportation."

Oh, I badly want to believe, I really do. But the future is seemingly forever far away. And when it finally gets here, it's going to cost us at least $130,000 and for what? As with so many other strange and exciting things—the gold rush, the Internet, Arnold Schwarzenegger—California is home to a lot of the cutting-edge companies that are still chasing the dream and developing science-fiction-y transportation options. And luckily for me, two of the most prominent of these companies are located within a day's drive of the sleepy cul-de-sac where my dad lives.

Trek Aerospace was founded in the late '90s in Sunnyvale, California, by an entrepreneurial engineer and former navy combat pilot named Mike Moshier. Within a couple of years, the government's Defense Advanced Research Projects Agency (DARPA) dropped nearly $5 million on Moshier and his skeleton staff to develop a duct fan–powered, vertically lifting flying machine for a single pilot. A bulkier, propellier jetpack for the twenty-first century.

Thus, the Springtail Exoskeleton Flying Vehicle (EFV) was born. Part minihelicopter, part Swiss rucksack, its two thirty-eight-inch-diameter fans loom above the pilot—who is strapped to the machine's tall base and standing on footrests like a knife thrower's lovely assistant—and connect to a gas-powered rotary engine. It is, in a perfect world, meant to be flown for two hours at 90 mph and as high as four hundred feet in the sky.

But it is not a perfect world. As of this writing, the Springtail EFV is still in the testing stage, hovering close to Earth for far less than two hours at a time. In 2003, in an effort to raise funding, Trek put the Springtail prototype up for sale on eBay, where bidding rocketed to more than $1 million in less than two days and eventually surpassed $7 million. But that's because the would-be Bondses doing the bidding thought it could fly. It was reposted with the understanding that it could not, indeed, be used, and the $50,000 minimum wasn't met.

When the DARPA contract expired in 2004, it was not renewed. Moshier has since left the company, and Trek has relocated to a tiny airplane hangar not far from Lodi Airport in north-central California. After a skin-cancer scare, Moshier started a new company called DermaFend, which manufactures sun-protection products, like the ExtremeShade cap and age-defying pills. He has clearly given up the dream. Still, if this is the jetpack of the twenty-first century, I need to see it up close.

And while I'm at it, I should probably push on an hour or so north to the University of California–Davis campus. That's where the mad genius that is Paul Moller has set up shop to construct what he hopes will be the first real-life flying car. The M400 Skycar, one of the most hyped pieces of technology to never take to the sky, has been featured in many of the country's larger media publications, and plenty of smaller ones, too, since the Canadian-born Moller first began tinkering some forty-five years and a reported $65 million ago.

The mechanical and aeronautical engineer has raised the dough through private investments, real estate deals, and brisk sales of his company's motorcycle muffler. In some ways you could say the money has been well spent, as Moller's people have put together quite a good-looking little number. The Skycar has the body of a dragster with four massive cylindrical engine casings bolted to the side. Each one of those houses two 360-horsepower ethanol-chugging rotary engines. The control system is made easyish by virtue of its microprocessor-enriched steering system. The Skycar is sleek and cherry red, and any self-respecting geek would definitely fork over at least half of his SpaceShipTwo ticket savings in order to own one. A Plexiglas cabin allows four passengers traveling at 300 mph to peer down at the beautifully undulating hills thirty-six thousand feet below. Or that's the dream, anyway. The Skycar is still yet to fly free of its steel safety line hooked to a crane.

When it comes to unrealized dreams in this department, Moller, a thrice-married former UC-Davis engineering professor who once built a flying saucer in his garage because he wanted to, is far from alone. The U.S. government has more than eighty flying-car patents on file— the Autoplane, the Planemobile, and the Airphibian are just a few of the wonderfully named crafts. According to the *Los Angeles Times,* of those eighty-plus inventions, only two have ever been certified by the feds for flight. And you can probably guess how many of those were green-lighted for mass production. That's right, none of them.

But, hey, I am going to be in the neighborhood, anyway. And in 2001 *Time* magazine had named both the Springtail (then called SoloTrek) and the Skycar two of its "Inventions of the Year." The story's writer noted, "Even in this age of space shuttles, supersonic jets and ultra-light airplanes, the quest to build the perfect personal flying machine still lures the world's inventors. Nobody is pursuing the dream of solo flight with more fervor than . . . Paul Moller and Michael Moshier." So there was that.

While I am on the West Coast there are also a few other guys I think I should stop in to see. In San Jose I'll try to meet up with aeronautics engineer Nino Amarena, with whom I had chatted briefly at the con-vention. Nino seemed keen on jet-engine technology, especially since, as he noted, "the shoe bomber," Richard Reid, had "given peroxide a bad name."

While in Los Angeles I hope to drop in on Nelson Tyler, who still works out of the same Van Nuys office in which he built his Olympian rocket belt. And then there is Juan Lozano. Larry Smith had described Juan as a rocket-belt "Buddha," to whom anyone who's ever dreamed of jetpacks must pay respect for spiritual enlightenment. And tequila. And maybe, just maybe, my first flight. For the Mexico trip, I'll con-vince my old friend Jeremy, now living in Los Angeles, to tag along as my camera crew.

So on a chilly mid-October morning, Catherine and I wake Oona before the sun comes up to catch our flight out west. In the taxi to the airport, I can hear Daphne gurgling in her car seat, while Oona sits wide-eyed and confused on Catherine's lap. A power ballad I'd previously thought of only as treacly and lame comes on the radio, but because we are a sleepy family in it together in this big, crazy world and because we are going to California, land of adventures, and because I am on a quest to find the most quixotic and brilliant and ridiculous thing in the world, because of all these things and more, my heart cleaves a little, and I find myself kind of liking the song—about faith, shame, love, and nudity from what I can gather—syrupy lyrics and all.

At the airport I buy a *New York Times* while Oona eats cold eggs. There's a front-page story on an eccentric group of rocket builders who meet every year in the desert to fire off their creations. Ky Michaelson is quoted as saying, "Most guys close their eyes and see women—I see rockets." He said the same thing at the convention! It's a good line, but c'mon. Anyway, the *Times* story gets me thinking, and I decide to write the paper of record my first opinion piece as soon as we settle in at my dad's place.

Our flight is mostly unremarkable, except for when Daphne somehow manages to barf *inside* my shoe and then, due to a related mix-up, Oona says very loudly and repeatedly, "Mommy, you have throw up in your mouth? Mommy, you have throw up in your mouth?" At which point we are laughing so hard we can't tell her not to yell that on a plane, or anywhere, really. I make the universal sign for "knock it off" by pretending to slash my own throat with an index finger, but Oona just looks at me blankly.

Until recently, I would say my dad and I were not especially close. But we've been creeping steadily in that direction of late. He moved from Baltimore to northern California when I was fourteen, and I've seen him infrequently since then, one or two short visits a year, tops. He is a psychiatrist by profession but also an information junkie and a hardcore dilettante who at one time or another has obsessively researched (among many other things) Sumo culture, bullfighting, perfume, basketball shoes, French literary theory, ballpoint pens, Roger Federer, and, his greatest love of all, physics. Many times I've wandered into Dad's study and found him reading textbooks on quantum physics. For fun.

When I first mentioned to Dad that I planned to seek out a working jetpack, he lit up like my daughter does at the mention of an airplane ride. Apparently, his latest area of interest is the history of flight; he has recently read three different books on the Wright brothers. Dad is so excited that he's already named the project "Orville" after half of the Wright brothers. (Orville was the one who watched his brother, Wilbur, die of typhoid fever in 1912.)

My dad moved to Oakland, California, in 1985, before settling in Walnut Creek and a life busy with projects and ideas. He has lived in the same snug single-story house since remarrying several years ago. He has spent much of that time working on something he calls the Brainmap, which looks like the world's busiest subway system and is meant, I think, to chart, in bright colors, the activity of neurons. He once started a literary magazine called *EDG*. When he was young he wore his hair long and frizzy, his sharp-boned face framed by bushy sideburns and thick glasses. In photos he looks like a geekier Peter Fonda. Today, his six-foot frame, once the sinewy home to a high school football lineman and an amateur boxer, has softened, and his hair has retreated into a silvery cloud puff. He used to dress only in white; now he wears all black—parachute pants, turtleneck, Nike hightops, military jacket—Johnny Cash as a navy SEAL.

The unifying theme of Dad's life, at least over the past two decades, has been his utter and complete devotion to the field of physics. He has missed holiday dinners and birthdays so that he might read more Feynman. He relaxes with Vincent Icke. For a man who has spent much of his life trying to make sense of it, this science might be as close as he'll ever get. In any case, to call Dad a riddle, wrapped in a mystery, inside an enigma is to suggest that if one were to only unwrap enough, he might eventually be better understood. I'm not sure this is true.

Soon after we started talking jetpacks Dad mailed me a four-inch-tall pilot with a 'pack and a nine-page operating manual along with an updated version of a nineteenth-century song he titled "The Daring Young Man and the Flying Jetpack," which began:

Once I was happy,
But now I'm forlorn,
Like an old jetpack
That is battered and worn,
Left in this wide world
To weep and to mourn,
Betrayed by mankind's oldest Dream.

The girls and I set up shop in my stepmom's bedroom. She and Dad keep different rooms, each claiming the other one's snoring awakens them. Once the porta-crib is erected, I sit down to write my *Times* op-ed piece. I didn't mean to begrudge these rocket men their ingenuity and happiness, I just thought *Times* readers would want to know a few things about rockets, jetpacks, and getting into the sky. Oh, and it seems like a good way to reach Richard Branson, too. My column begins:

The front page of last Saturday's *Times* carried a lengthy story about 'backyard rocketeers.' The piece described a group of about 100 men—and yes, everyone mentioned was male—who spend far too much time tinkering away in their garages and are known collectively as the Tripoli Rocketry Association. The men meet annually in the Black Rock Desert of Nevada to launch the rockets they've built themselves—sometimes literally in their own backyards—94,000 feet into the air.

I'm sure most people read the piece with bemusement and delight. Here was do-it-yourself innovation at its finest. The story struck a chord. The next day it held steady at number 7 on the Most

E-mailed List. But I believe that certain readers took in the piece with pleasure as well as with a pang of dismay. That's because, while the story mentioned the rocketeers' ability to launch Weber grills, porta-potties, bowling balls, and pink flamingos into the stratosphere, there was one glaring omission from that lineup: people.

It went on from there, outlining the Rocket Belt history at Bell, the convention, and the cloudy jetpack future. I recalled what Suitor had said in Niagara Falls: "So if anyone here knows Bill Gates, call him up. Get him to write a check for $1 million."

"It's a good thought," I write.

But rather than Gates, I might call on his billionaire buddies, Paul Allen and Richard Branson. In last Saturday's *Times* story, it was noted that both Mr. Allen and Mr. Branson have contributed heavily to the space tourism and space entrepreneurship movements, best typified by the increasingly popular X Prize. Or I might call on Budget Suites owner Robert Bigelow, who has invested $500 million in building—wait for it—an inflatable space hotel.

For those dynamic dreamers still pursuing the far more practi-cal, more proletariat, and, let's face it, far cooler jetpack, there can be few words less galling than "inflatable space hotel." Because if Bill Suitor is right—and no one knows this stuff better than Suitor—it would require much less than $500 million for Will Breaden-Madden to make his ShamRocket available to the masses.

So how about it Messrs. Allen, Branson, and Bigelow? Why keep futzing with daffy ideas about blow-up space motels when you can make history here on Earth? For 1/500 the investment, you could re-ally change the world. Because apparently, the future is, once again, now. At September's convention, Mr. Suitor quoted his mentor Wen-dell Moore, who once said that jetpacks "are an idea 50 years ahead of its time." Mr. Suitor then added, "Well, it's been 49 years."

I fire it off and the paper's editors are apparently so enamored of my work, they decide it's too good to sully with anything as crass as publication.

At eight the next morning, Dad and I climb into his white pickup truck. We are off to see Nino Amarena at the Hiller Aviation Museum in San Carlos, a booming townlet just south of San Francisco. The museum was founded in 1998 by local helicopter innovator and entrepreneur Stanley Hiller Jr. Though perhaps best known for creating the world's first coaxial helicopter, in certain circles he is remembered as the man who gave us the Hiller Flying Platform, a kind of oversized duct fan–powered manhole. (The museum's current VP is a tall, slender fellow by the name of Willie Turner—*Rocketeer* obsessives might recognize the name, as Turner did some of the stunt-plane flying in the flick.)

Nino has built something he calls the ThunderPack, which resembles the Bell Rocket Belt but can reportedly fly for up to ninety seconds. I want to find out how Nino has done it and also to see if that means he might be developing an even more ambitious flying machine. So far, his ThunderPack has generated enough interest that a Japanese technology company is currently analyzing the machine to determine if it can up the air time even more, and in the process make the 'pack a viable earthquake-disaster rescue option. Nino told me on the phone that a nondisclosure agreement he's signed means he cannot discuss the Japanese involvement at this time.

It's my idea to have Dad along on a couple of these research and development missions. His expertise in physics will surely help the cause, and it will give us a chance to catch up in the curious, occasional, meandering manner that we do.

There is much to try to figure out about him, after all. One of the many sides to his hexagonal personality is a decidedly tough-guy streak, a quality he attempted to impart on his two boys growing up. When playing all-time quarterback for my older brother, Asher, and me during one-on-one tackle-football games, his favorite play call was the dreaded "fullback slam." This was a handoff, straight up the gut. Right into the other guy's sternum. That might seem cruel, but I can kind of see his thinking on this one. Boys are going to beat the crap out of each other anyway, so why not do it in a constructive way that involves the outdoors and exercise? To this day, Dad mightily enjoys a round or two of kitchen boxing, wherein he tiptoes up, surprisingly quickly, for a bit of open-hands cuffing at my ears as I retreat and howl for mercy.

After driving a tiny Honda hatchback for the better part of two decades, it's good to see him behind the wheel of this truck. It suits him—he's a still-rugged guy with a vivid pioneering and adventuring streak. Not sure if it's an apocryphal tale, but he claims to have hopped freight trains across the country soon after his eighteenth birthday. The image of him, swinging along the rails with packs of soot-faced hobos, thrilled me as a kid.

On the rearview mirror hangs a dream catcher; the bumper gleams with a gold NRA sticker. I still don't know what to think of this latest obsession, but somehow he has made the unlikely transition from consumer of French-lit theory and devotee of Derrida to being a ravenous information sponge of all things gauges, hammers, calibers, and shotgun sights. He frequents a shooting range and possesses a small arsenal of weapons, which I've asked him to hide in the garage while the girls are around. If I had to guess, I'd say the guns are part of a larger transformation from East Coast liberal hippie to West Coast die-hard libertarian.

He turns the key in the ignition. "Flight 11094595 requesting permission for takeoff," Dad barks in perfect imitation of a Hollywood pilot, subbing his address plus ZIP code for the flight number. He backs out of the driveway, and the commuter cup I've precariously placed on the dashboard flies into my lap, hot coffee suddenly bubbling unfortunately on the crotch of my jeans. We cackle at the auspiciousness—and we're off.

It's one of those ad-copy California fall days with a brilliant blue sky and shimmering sun. Soon we are in heavy traffic, approaching the San Francisco Bay. Dad has prepared a seventeen-page printout for me that he's titled "Flying Fizziks." It's a quickie reference guide to help get me up to speed on several aspects of this airborne business. The document opens with a list of symbols and terms like *force, acceleration, mass, wing area, thrust, drag,* and so forth. Later, a section called "The Basic Basics" includes entries like "POWER: 1 horsepower = 550 foot poundals/second." There are a handful of charts and diagrams and subsections such as "Dimensional Analysis, Wing Load and Carrying Capacity." The definition for yet another section, called "Functions," begins, "Example: You are at an orgy. In one room are all the women. In another room are all the men. Each man is married to a particular woman, and vice versa. The rule is: each man is to go to the room where the women are and pick his own wife. This is a very boring orgy, but a good definition of a function: A function is a way of connecting things in one group uniquely to things in another group."

It's a lot to take in. When on earth did he have time to compile all of this? After I've flipped through the pages for a few minutes, Dad begins quizzing me, "Do you know why a ballistic missile is called ballistic?"

I have to admit that I don't and in doing so am immediately six years old again. When Dad was in medical school back in Maryland, back before we were all dispersed around the country, he used to prep for tests by quizzing Asher and me. "So, what's the latest on narcoleptic cataplexy?" he might ask. Blank stares. Silence and confusion. Thoughts of Wiffle ball. Undeterred, he'd press on. "What? What's that you say? There's a new drug, huh? Cool. And it's called? What is it called?" More silence. "Oh, yes, I think you are right. I think it is in fact called sodium oxybate. Okay, yeah, I get it. Sodium oxybate. Thanks."

Now we're in the truck, talking ballistics. "That means its destination is preset," he tells me, eyes on the tollbooth. "Everything's been arranged. There's no control."

I stare groggily out the window, the coffee on my crotch almost dry. It was another rough night sharing a room with Oona and the Daph, another night of desperate resting between bleats. I'm already fading. My eyes are two burning coals. An old dude with a graying ponytail is riding a motorcycle next to Dad's truck. He's wearing a leather jacket with an American flag patch on the back that reads, "Try burning this one asshole!"

"The two big problems in flying, like in any movement, are, one, getting the power to make the movement happen and, two, steering the movement."

He makes it sound simple. Soon we're over the Bay, and Dad is discussing the weight distribution of rockets, I think.

"There's been a ballistic system in place for thousands of years. Any idea what that is?"

"A catapult," I flat-out guess.

"Well, that's a ballistic system but. . . ."

Trying to think of something clever, or at least something that won't embarrass me, I reply, "A bow and arrow."

"Right!" Dad is beaming. Then he makes a leap and loses me when he concludes, "So with jetpacks, it's not really flying—more like an extended takeoff and landing." I get the feeling there is a deeper significance at work here, but I'm too exhausted to grasp it.

A few minutes later we're passing the It's It ice cream sandwich factory, closing in on San Carlos. Dad is talking about $e = MC^2$. I hear the words *fusion* and *fusion reactions*.

"Which are what exactly?" The warm sun is whispering to me about sleep.

"The method by which something releases energy—in this case we're talking about force and hydrogen molecules."

We are almost there—in fact, there is a sign for the museum exit now. Looks like we will be just on time, but instead Dad is frantically trying to figure out if that is in fact the right exit and we . . . sail right on by.

"Fuck, fuck, fuck, fuck, fuck," he vents through clenched teeth, jabbing a foot on the gas pedal. A minute later, rising to the thrill of the chase, he takes the next exit, whips a U-turn across the overpass, and zooms back. Now we are in danger of being late, and Dad is flooring it as we careen off the freeway and onto the quiet nearby streets of the suburbs. My father checks the time on his compass-equipped black wristwatch and guns the pickup's engine. He's a bohemian Steve McQueen. We squeal hard around a corner, fly past a Burger King with a helicopter model in the playground, and enter the museum's parking lot. Dad hops a curb, the truck's shocked hydraulics bounding us along, until he swerves into a spot, miraculously not grazing a cherry-red Mercedes convertible with the vanity plate "Ninetto"—Little Nino. "We're here!" the shrink at the wheel yelps, giggling into his black turtleneck.

I unlock my tensed shoulders, glance nervously at the front door, and see a suave, sophisticated man with a middleweight's build, wearing a floral-patterned tie and the neatly trimmed hairstyle of a New Wave filmmaker, talking calmly on his cell phone. Nino Amarena. I'm just glad his car is still in one piece.

"Let's . . . give it a minute before we get out," I suggest.

If Nino knows we nearly smashed his beautiful car to bits and is perturbed, he doesn't let on. We shake hands, I introduce my dad, and Nino leads us to a conference room in the back of the museum. One wall is lined with a kind of "Great Men of West Coast Engineering" hall of fame—Cliff Henderson, Bill Barber, Wayne Handley. Men with buzz cuts and impressive chins. Mathematical minds.

We sit at one end of a long lacquered table. I'm at the head, facing a broad window, beyond which is the museum's lobby. Nino is wearing a tiepin commemorating Alan Shepard's Mercury flight of 1961. As he unfolds the decadelong saga of his ThunderPack, I look through the window and watch mothers chase toddlers too young to dream big.

Carmelo Amarena was born in 1957 in Argentina. Like many Argentineans, as a boy he was enchanted by the five-time Formula 1 racing legend Juan Manuel Fangio. And like Fangio, Nino was born to Italian parents in Buenos Aires. It would turn out that the two men had other qualities in common as well, like a love of speed and the desire to continually look for the next challenge, the next rival.

Growing up, Nino was also obsessed with a Japanese television show called *Tetsuwan Atom* (Mighty Atom). A kind of protoanime cartoon series, in English translations, *Mighty Atom* starred Astro Boy, a preadolescent robot with slicked-back black hair, a metallic bathing suit, and jets where little legs should be. As the story goes, Astro Boy was built by a kindly professor, Pinocchio-style, after his own son died in a car accident.

These twin obsessions lodged in Nino's consciousness, and many years later they manifested themselves in the form of a master's degree in electromechanical engineering and a bachelors in electrical engineering. Soon after graduating from college, Nino moved to Palo Alto, fell in love, got married, raised a son, settled down, and embraced the California good life.

While working with a racing team in the mid-'80s, Nino first encountered such things as wireless LAN systems and WiFi technologies. Around this time, he also briefly met his idol. A friend of his who was a professional driver often stayed with Nino when he raced in the area. During one of those trips, Nino's friend arranged a meeting with his Formula 1 racing hero, Fangio, who was visiting one of his sponsors. "It was like a childhood dream," Nino tells me and Dad, practically climbing out of his seat with excitement. "It was like meeting Neil Armstrong."

The engineer's interests and capacity for work began to expand. His reputation grew. He became involved with the microsatellite Clementine 2 project, which he tells us is "an asteroid interceptor, in case we have one of those Armageddon situations." The purpose of Clementine 2, he says, "was to prove the theory of acquiring a target and launching a kind of offensive to an asteroid that could crash into Earth. The speed of an asteroid is 50,000 mph, so it has to acquire the target within three seconds."

That work got Nino interested in and thinking about propulsion, and he soon met a group from the Georgia Tech Research Institute, whom he assisted on the development of a mechanical flying bug the size of a boomerang. With wings made from old Coca-Cola cans and a roachlike appearance, it was dubbed the "Coke roach." Later the machine was renamed the Entomopter for its resemblance to a helicopter using entomological flight control. Nino tells us that the Entomopter is equipped with a tiny camera and can be used for surveillance in hard-to-reach areas. He says he believes that the machine will be flying on Mars by 2010. The Entomopter is powered, primarily, by hydrogen peroxide. "So that's how I got involved with hydrogen peroxide," Nino concludes the first chapter of his life. "It was micropropulsion."

In 1996 Nino read an article in *Smithsonian* magazine about the history of Bell's Rocket Belt that changed everything. "Being born in South America, you can drive any speed you want, just like in Germany. One of my biggest pet peeves in the United States, particularly in California, has been the speed limit. From here in San Carlos to where I normally work in Sunnyvale is twelve miles, and it takes me fifty minutes. So I was always looking for an alternative way to commute to work." The Rocket Belt gave him an idea. "That's how the whole thing, really, like in the nut of a joke, started. The idea was to build something that was utilitarian."

It may have started as a joke, but for the next five years Nino spent every free moment working to beat Bell's twenty-one-second flight time.

With funding from a group of investors, including racing team owner John Della Penna and his brother Horace, who provided the cars for the Sylvester Stallone film *Driven,* the Argentinean set to work. Though he admired the man greatly, he was not interested in re-creating Wendell Moore's machine. He wanted to build something that could be used for more than mere exhibitions, envisioning earthquake-rescue missions, flood assistance, and perhaps even a better commute.

To increase the amount of time he could stay in the air, Nino knew he needed to make the pack lighter and the fuel more efficient. He opted for a bipropellant system where the machine either could operate on pure hydrogen peroxide or could combine two types of fuel to increase its specific impulse, or Isp. The Isp indicates how powerful a fuel is compared to how heavy it is. The higher the Isp, the more potent it is. Whereas Bell's Rocket Belt fuel had an Isp of 122, for example, the space shuttle Columbia used hydrazine, which has an Isp of 215. "That's one of the reasons they told us not to touch any of the debris when the Columbia crashed in 2003," Nino tells us. "Because of the hydrazine—it's carcinogenic."

So hydrazine was clearly out as the second fuel type for Nino's 'pack. He ultimately settled on using hydrocarbon as the second element. That's the same stuff used in kerosene, paint thinner, and liquid wax. "All you have to do is inject the peroxide first, and then after a certain amount of time you inject the hydrocarbon, and if the flash point of that hydrocarbon was below 1,285 degrees Fahrenheit, it would ignite automatically. So you had a very elegant way of turning it on and turning it off without having that horrendous problem that plagued us during the earlier stages of the space race, where rockets would go half a foot and then explode."

Dad looks pleased with Nino's work. "Goddard knew that," he offers, conjuring the imaginative, difficult rocketing genius.

"Goddard learned that, but he was very secretive," Nino counters. "You know when Lindbergh gave him money and everything, trying to

help? He was just very, very tight-lipped, probably because of the problems he had. That's why he had to move to Arizona or Colorado...."

"To New Mexico," Dad corrects.

"He had some problems—he had an explosion, and I guess they
kicked him out of where he was living..."

"On his aunt Essie's farm."

I steer the conversation back to Nino and his ThunderPack. By
1998 he had completed four design phases and had solved his bipropellant fuel problem. He had met Bill Suitor and spoken with a bunch
of old Bell engineers and guys from NASA's jet-propulsion laboratory
in Pasadena. He believed he could come up with a new breed of
rocket belt that could fly for ninety seconds. "Still, that's not going to
make people scream with joy," Nino says, neither screaming nor looking very joyful.

So he contacted Sam Williams and his son to ask about jets.
Through them he learned that no one was manufacturing jet engines at the time that were both small and light enough and sufficiently powerful to do the job. Nino believed the drones used by the
military for target practice might work, but they were extremely expensive, about two hundred thousand bucks a pop, and their dura-
bility was unknown. "So that was the initial intent, to build a
jetpack, and it never died, even throughout all the years of building
the 'pack." Nino looks a little defeated. He tells us that today GE
makes a good and light jet engine but that to start over and build a
jetpack prototype, even for a man of his cunning and enthusiasm,
would run about a half million dollars. "The problem is now the
funding has run out."

That's not the only problem. Nino has had a devil of a time acquiring fuel. When he was ready to begin testing his engine in early 2001
there were only two companies that he knew of manufacturing rocket-
grade peroxide. He contacted BECO-FMC in Pasadena, Texas, but they
couldn't ship it to him because they claimed the required drums, sen-

sitively built from 99.5 percent aluminum, each cost about eight thousand dollars to make. "That's about as soft as this," Nino says, squeezing my notebook.

"I said, 'We can go to the moon, but you can't build a freakin' drum!'"

That seemed to be the case. So Nino spent a year constructing the drums himself, ultimately shipping about 115 of them to Pasadena, Texas. "But then came September 11th, and all hell broke loose."

BECO-FMC eventually shipped Nino four drums of hydrogen peroxide at 90 percent strength. In those containers was stored 1,200 pounds of fuel, enough to fly his creation while secured to the earth by a steel cable a total of twenty flights for thirty seconds each.

Before strapping the ThunderPack on himself, Nino tested the machine's strength by affixing to it a dummy of equal weight he'd named Mario. "I just have this tendency to give names to things, like my hummingbird feeder is called 'Peekie,' because in Spanish *pickafluer* means 'hummingbird.'" Nino giggles.

Mario soared without incident, and then it was Nino's turn. One day in the summer of 2001, he drove the ThunderPack out to an old World War II air force target-practice site in Patterson, California. "I had never heard a rocket of 300 pounds of thrust, 1,280 jet horsepower, about thirty-six inches away from me going full blast."

Dad and I look at one another—no, neither have we.

"When you tested on bipropellant mode, it makes a lot more noise than monopropellant mode. So one of the guys that were at the test said it sounded like thunder, and I said, 'That's it! We're going to call it the ThunderPack.' Even though it was very deafening, it was very exciting. I could feel my heart trembling."

I ask Nino what it was like to fly his creation, to become a wingless, soaring Dr. Frankenstein.

"I'll tell you one thing, do you remember the first time you kissed a girl? You had the butterflies in your stomach. The first time you rode a bicycle—do you remember that? It was a combination of exhilaration and fear and adrenaline. And of course at the end of the test there was, 'Wow, did you hear that!' It's almost like you burn the adrenaline through your lungs to get it out of your system."

He flew for ninety seconds, hovering about a foot off the ground. Two weeks later he rose to nearly three feet above Earth. "It was scary. Because you are so concentrated on what you are doing, and not only are you the pilot, you are the guy who designed it. You always have doubts, right? Did I weld that thing properly? Is that strong enough? Will it hold? Four-inch bolts holding the whole thing. Once you give the thrust, all those things disappear. Now you are paying attention to 'wooooow,' because it feels like someone is hugging you from here, someone is lifting you from here."

Nino tugs at the back of his shirt collar. He has an expression of wonder. We all do. The morning has given way to early afternoon, and the shadows are plunging on the long wall of plaques. It's almost time for our museum tour, but there is one last thing I'd like to ask: "What can you tell me about the throttle valve?"

Nino lights up. "My valve is not the same. It's a dual valve; it's a different flow. Because I'm using bipropellant *and* monopropellant, even in the monopropellant mode the valve is different than the regular rocket packs because it puts out different mixture ratios, so I had to find a happy point between monopropellant and bipropellant which is totally different than a true, strict monopropellant valve, which is what all the other 'packs out there use."

Nino reflects on this.

"A lot of people I talked with at the convention were asking questions that made me think they think there is some voodoo going on with this. Like, why is the throttle so finicky? Why does it have to be so precise?"

Of everyone I've met so far on my quest, Nino Amarena seems like the most qualified to build us all jetpacks. Yet it appears that his best shot has passed him by. Now the most he can hope for is a rescue vehicle jointly produced with a mysterious Japanese collaborator. Given that the contract on the project is up in about a year, even that seems to be a long shot.

When he had the funding and the drive, he lacked the equipment. Now his project is out of cash, but with half a million bucks he believes he could very handily throw together a jetpack prototype, and I have no reason to doubt him.

We take a quick spin through the museum. Nino guides us, while Dad and I shuffle behind. We pause under a model of a man attached to a hang glider reminiscent of the one the great German inventor Otto Lilienthal tested in the 1890s. Lilienthal was a brave and ingenious airman until one of his monoplanes stalled, bringing him crashing to the ground and an early demise in 1896. A sparkling biplane pusher called *The Diamond for the Mines* of Pittsburg, California, hangs from the ceiling—this is one of the models Nino, as a museum curator, has helped refurbish.

Then we are standing in front of the Wright brothers' motor, kept under glass like a rare jewel, which was used in 1911 for the first flight from the West Coast to the East. Its darkened innards coil, taking on the shape of an exotic amphibian. Dad sighs deeply. "It's exquisite, isn't it?" he asks.

"Yes," says Nino quietly. "It is."

We climb inside a mock-up of the Condor, Boeing's spy plane from the '80s. It's impressively spacious—about as big as a two-bedroom apartment in Brooklyn—and the cockpit is stuffed with more glowing instruments than I can count. Although it is certainly an amazing feat of physics—how do you begin to navigate something with a two hundred–foot wingspan?—it strikes me as sort of the antijetpack. Too much apparatus, too complicated. I quickly deplane.

Next we move to a corner of the museum dedicated to its founder, Samuel Hiller Jr. Several frying pans are displayed in one glass case.

For a moment this strikes me as the quirkiest flying thing yet and not all that scientifically advanced, but then I remember that's how old Sam got his start, in the kitchenware business. If it weren't for those fast-selling frying pans, the world may never had seen his Flying Platform, pictures of which we are ogling now.

The Hiller Platform was first tested in 1955. The pilot stood on an eight-foot-diameter fiberglass disc, looking like a game show host of the future, riding a piece of the stage that's broken loose to wander the sky. The platform, a vertical takeoff and landing device, or VTOL, was powered by three 44-horsepower, four-cylinder engines and two counter-rotating propellers, or duct fans. So it's something of a precursor to Trek's Springtail and even Moller's Skycar. A flight suit–encased, crash-helmeted pilot captained the 180-pound Hiller Platform by engaging the engines and fans and then steering the massive manhole through kinesthetic movements. That is to say, by squirming.

Tests commenced, and Hiller's gang proved the platform could climb a few feet in the air. In one memorable demo, a soldier managed to hover above the ground and fire his rifle without falling. But as the Wright brothers experienced many years earlier, a machine that is controlled by a contorting pilot hasn't much of a future.

Unlike with Orville and Wilbur, this proved Hiller's downfall, as the army ultimately considered the platform too unstable to reliably transport a single soldier. Around this time, Bell was making great progress with its Huey helicopter, and the military turned its attention and funding eastward to Buffalo. While gazing at the sepia-toned images, however, the strong-chinned pilot riding the air as if by magic, it is easy to see why the Hiller Platform once looked like a snapshot of the future zapped back at mid-'50s America.

Upstairs I wander away from Dad and Nino, and begin idly reading the short biographies attached to first one and then another flight obsessive. I stop short in front of Bessie Coleman, examining with my mouth hanging open in wonder the short sketch of her life. Coleman was the tenth of thirteen children, born in 1892 to sharecropping parents. She grew up poor in a small Texas town. From an early age Coleman dreamed of flying. The dreams were so intense and so enduring that she eventually took French language classes in Chicago so that she might get her pilot license in Paris. American schools, at the time, would not accept her because of her race and gender. She would go on to become the first African American woman pilot. Through the early 1920s, Coleman gained a passionate following at air shows all over the country. But in 1926, twelve minutes into a flight from Dallas to Jacksonville, Florida, Coleman's aircraft went into a severe nosedive; she couldn't pull out, and Coleman fell from the plane about five hundred feet above the ground. Her funeral in Chicago drew a crowd of ten thousand people. By the time I finish reading the brief bio, I find myself choking up over Coleman's unbridled passion for the sky.

Then it's time to say good-bye to Nino. We are in the gift shop, hastily making plans to stay in touch. He promises to keep me posted on what happens with the Japanese company. He'll also get me a good contact number for Richard Branson. Sir Richard's office, it seems, once called Nino inquiring after this ThunderPack. This is the first time during my travels someone has offered help in reaching the world's most famous Virgin, but it won't be the last. Nino shakes my hand and disappears, passing a stack of toy wooden gliders on the way. (I might not have guessed this at that moment, but in late 2007, Nino invited Bill Suitor out to test his newly revamped ThunderPack. The video of Suitor's twenty-four-second test flight soon flooded the Internet, and Nino was inspired enough to offer working models for a hundred thousand dollars each.)

I **buy** Oona a wind-up StratoCopter plastic plane and a Hydrotech Aquazone Aqua Launch Water Rocket. Dad and I head to the Sky Kitchen Café at the adjacent flight school for lunch. Over burgers we marvel at Nino's wisdom, courage, drive, and haircut. Above us, models of World War II fighter planes dangle by thin strings from the diner ceiling. A small lunch crowd mumbles beneath the whirring and whining of a handful of prop planes circling the pavement just beyond the restaurant's back window. I watch the planes, mesmerized. The pilots, regular guys like me, are about to soar, to drift soundlessly over the loveliness of mid-California's rusty gold surface.

"So," I start, not sure why I am getting into this, "you'll be sixty in two days."

"Yep." Dad takes a big bite of burger, mayo streaking his cheek. I suddenly recall spending much of my childhood contemplating the food stuck to my dad's face.

The idea, the age, my nonquestion for him hang in the air until I can't tell if he's still thinking about it.

"But you know," he starts slowly, gazing out the window, "I like to live like Goethe said—live like you're going to live forever, and hopefully you'll be able to do something you love."

Two hours later we're in the backyard in Walnut Creek. It's dominated by sprawling ivy patches, a low-slung hammock, and an oak tree looming over the concrete patio. I'm winding and winding and winding the rubber band of the StratoCopter engine while Oona looks up at me with enormous, expectant eyes.

"Okkkkaaaayyyy," I say, drawing out the drama. Oona stands on the tips of her tiny toes, rocking friskily side to side in anticipation, like a blonde, shoeless R2-D2.

I let go of the propeller, and . . . the StratoCopter nose-dives for pavement. Oona looks defeated. I grab the thing and quickly wind and wind and wind. "Minor setback, sweetie. This time we're golden. Okay, here we go, one, two . . . threeeeeeeeee!"

The StratoCopter whizzes straight down. My daughter is losing interest. She'd rather play with the tennis ball and Jacques, my stepmom's bichon frise, named for Derrida, as you might've guessed.

"One last time—here we go." Wind, wind, wind, wind, wind, wind, wind. And release. The StratoCopter soars, glides, glistens in the perfect California sunlight. Oona's eyes go even wider with delight. Ah, flight, you magical, mysterious beast.

Incredibly, the simple little machine climbs ten, twenty, twenty-five feet into the air—where it gets snagged by a tree branch and dangles by a plastic wing, taunting us. "That's okay, that's okay, check this out." Perhaps only mechanical engineers are capable of thinking more quickly on their feet than parents. I grab a tennis ball and chuck it at the StratoCopter, missing by inches. Again, same result. And one more time. Finally, on the fourth try, I peg it, and the miserable contraption falls to Earth. Oona shrieks gleefully—finally, the damn thing has thrilled her. "Any interest in the Hydrotech Aquazone Aqua Launch Water Rocket?" I ask.

Two days later, Dad and I again climb into his pickup just after dawn. We are going to hit Trek Aerospace in the morning and see Moller's Skycar in the afternoon. We zip along Interstate 680 out of town, and soon we've left the exburbs behind for the smaller roads with bigger numbers cutting through wide pastures.

Dad has spent the morning preparing, and he hands me two big binders full of paper printed from various Web sites. One is the operator's manual for Trek's OVIWUN, a remote-controlled, pilotless version of the Springtail, about the size and shape of a metronome with miniature duct fans affixed to its top. This is meant to be a police or military surveillance tool. The other book is a three-ring binder he's titled "Projects for Personal Flight," under which it is attributed to EDG Design Group—my father's company of one.

In this binder he's organized more information on the history and products at both Trek Aerospace and Moller International. There is also material on the autogyro, which begins: "An autogyro is a type of rotary wing aircraft supported in flight by lift provided by a rotor.

Unlike a helicopter, the rotor of an autogyro is driven by aerodynamic forces alone once it's in flight, and thrust is provided by an engine-powered propeller similar to that of a fixed-wing aircraft." And then continues: "The autogyro was invented by Juan de la Cierva y Codor-níu in 1919, and it made its first successful flight in January 1923 at Cuatro Vientos Airfield in Madrid, Spain. De la Cierva's pioneering rotor technology paved the way for the invention of the helicopter 13 years later."

It's clear that Dad has a serious crush on the autogyro, the latest in a life full of them from a man with an immense capacity for deep and mysterious feelings. Before Mom, the crushes—or at least great interest—came fast and furious on many women, then came Mom, and then the mistresses and the girlfriends, science, philosophy, and handguns and Indian motorcycles, and now, also somewhat unexpectedly, my daughters.

Before Oona was born he'd told me, a bit awkwardly I thought, to never have kids, that he never intended to be a grandfather. But her birth had a transformative effect on Dad. The first inkling of this came when he debated over what she should call him. Grandpa would never do, but maybe Tex? He liked Tex a lot, and it was looking like Tex until a dark-horse candidate overtook Tex in the home stretch—Papa Jam. Papa Jam spent the year between Oona's first and second birthdays mailing, like clockwork, one letter every week to her. Here, a short sample from the beginning of one of them:

> Hello Oona Toona,
> I am sure you will not believe, as I cannot believe, the things I have had to avoid this past week. On Monday afternoon there was a mosquito, a very big mosquito, in my study. How big? Imagine an eclipse of the sun with a wild whine like a deranged Cuisinart.

My daughter was fifteen months when she received this missive.

Perhaps the coolest movie character ever—*Star Wars* creator George Lucas's bounty-hunting Boba Fett.

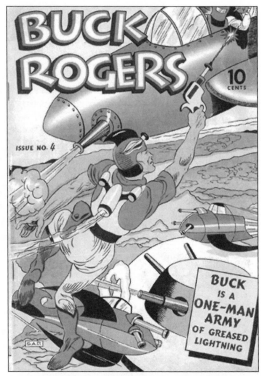

The guy who started it all: Buck Rogers and his dainty looking but powerful 'pack.

When box office take is adjusted for inflation, 1965's *Thunderball* remains the highest-grossing James Bond film of all time. Coincidence?

Lost in Space, with stars June Lockhart and Guy Williams, is a pop culture touchstone for jetpack obsessives the world over.

Photo by Gabi Rona, courtesy of the Motion Picture and Television Photo Archive

In the early 1960s, Bell Aerospace's Wendell Moore looked into the future—and saw his flying creation replacing cars.

Courtesy of Kathleen Lennon Clough

A Toronto newspaper story imagined how Wendell Moore's invention could change the future for commuters. That's Hal Graham in the air and Moore's daughter and son playing Graham's family.

Courtesy of Hal Graham

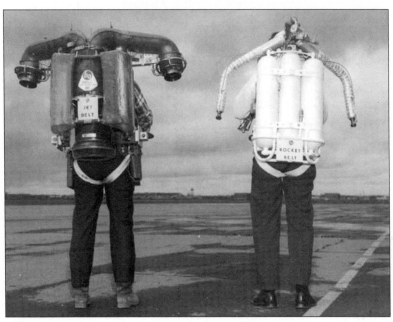

Bell's Rocket Belt (right) and bulkier Jet Belt. Brief flight times, heft, and cost ultimately doomed both inventions' chances with the military.

Courtesy of Kathleen Lennon Clough

At a 1961 Bell demonstration for the Pentagon, Harold Graham wowed the crowed on a packed lawn.
Courtesy of Hal Graham

Does that silhouette look familiar? Graham salutes President Kennedy in the fall of 1961 at a Fort Bragg demo.
Courtesy of Hal Graham

The Bell gang liked to attach rockets to almost anything—even this everyday office chair.
Courtesy of Kathleen Lennon Clough

At a 1961 Bell demonstration for the Pentagon, Harold Graham wowed the crowed on a packed lawn.

Courtesy of Hal Graham

Does that silhouette look familiar? Graham salutes President Kennedy in the fall of 1961 at a Fort Bragg demo.

Courtesy of Hal Graham

In 1964 alone, Bell pilots flew at nearly 400 exhibitions including in Honolulu, Copenhagen, London, Ontario, Sydney, and 170 times at the New York World's Fair.

Courtesy of the Daily News

(NEWS foto by Hal Mathewson)

Spaceman Robert Courter circumnavigates the Unisphere at the Fair.

Rocket Man Flies at Fair

A giant leap for mannequin kind—at the Niagara Aerospace Museum for the First International Rocketbelt Convention.

Photo by Joanna Ebenstein

With government backing, in 1958 New Jersey's Thiokol Chemical Corporation developed
something it called "Operation Grasshopper."

Courtesy of Mark Wells

The Bell gang liked to attach rockets to almost anything—even this everyday office chair.

Courtesy of Kathleen Lennon Clough

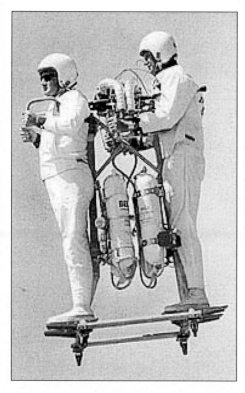

Testing the two-man POGO, which Bell hoped would revolutionize lunar transportation.
Courtesy of Kathleen Lennon Clough

Bill Suitor leaps out over the field at the Los Angeles Coliseum, flying Nelson Tyler's belt for the opening ceremonies of the 1984 Olympic Games. Suitor's seventeen-second swoon was seen by 2.5 billion television viewers around the world; it is perhaps the most famous flight of its kind.
Courtesy of Nelson Tyler

"Mountain hopping, it's sort of the jet age answer to mountain climbing."

"All you need is a rocket pack, a pretty assistant—and you're ready to hop your first mountain. Our take-off point, Cheakamus Canyon, in the Tantalus Mountains of British Columbia. Susan makes a last-minute check, setting her stopwatch to my fuel gauge.

"Now the last and perhaps most important piece of equipment—my radio helmet. Without it, Susan can't signal me to land when my fuel starts getting low.

"3...2...1. Varooom! Suddenly I was on my way skyward. I felt like a giant bird who could soar to the top of any mountain. All I could think of was...next hop Mt. Everest.

"Later, we joined a friend in the comfort of a mountain inn. And toasted our adventure with Canadian Club." It seems wherever you go, C.C. welcomes you. More people appreciate its gentle manners. The pleasing way it behaves in mixed company. They admire its unmistakable character. A taste not matched by any whisky, anywhere. Canadian Club— "The Best In The House"® in 87 lands—

Canadian Club
Imported in bottle from Canada

Job No. CC-679
Magazines, 1971—Page, Color (9 3-8 x 12 1-8 inches)
Proof 2   V&H 11745   8-19-71
Advertisement prepared by LaRoche, McCaffrey & McCall

Nelson & Susan Summer 1971

A smashing success: Hollywood inventor Nelson Tyler shows off his goods in a 1971 whiskey ad.
Courtesy of Nelson Tyler

Tyler Rocket Belt
worlds smallest personal flying machine
fueled, ready for flight

Nelson Tyler's son, Scott, models dad's latest invention.

Courtesy of Nelson Tyler

He may have looked like a "hood ornament," but the Rocketeer has inspired plenty of garage tinkerers to dream big and to attempt to build their own jetpacks.

Courtesy of Dark Horse Comics

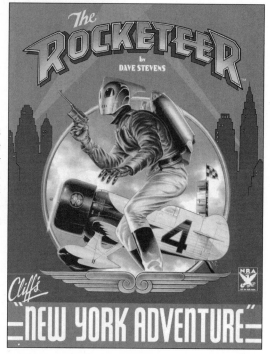

In 2006, Isabel Lozano became the first woman to ever fly a rocket belt (on a tether or otherwise), when she cranked up the machine her father, Juan, had built for her.

Courtesy of Juan Lozano

The closest I've come so far—that's me trying on Juan Lozano's merchandise.

Photo by Jeremy Kasten

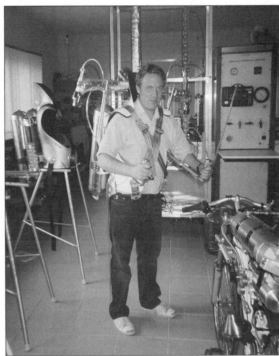

License to Il: Thanks to the magic of Photoshop, even North Korean dictators share in the dream.

Courtesy of Associated Press

A 21st-Century Man, Go Fast! energy drink's Eric Scott flies near Wendell Moore's old testing grounds, in front of the Niagara Aerospace Museum.

Photos by Joanna Ebenstein

Diddy or didn't he? Sean Combs allegedly traveled by jetpack to a press conference for the 2005 Video Music Awards, for which he served as host. He is far from the only celebrity pining for a 'pack.

Photo by Larry Marano

New Jersey's Gerard Martowlis has spent eight years and about $40,000 building his rocket belt. In his backyard in early 2007, he tested it for the first time.

Courtesy of Gerard Martowlis

We pull off Highway 12 onto a narrow road squeezed between densely packed rows of corn. Soon we are whizzing past acre after acre of Lodi vineyards. Dad turns left onto a bumpy, gravelly, desolate stretch of pavement lined with airplane hangars of various sizes. We bounce into Lodi Airport's Center Hangar 21—home to Trek Aerospace. "It looks a lot like the Wright brothers' shop in Kitty Hawk," Dad says. Hangar 21 is a low, squat building, about as big as a high school gymnasium, made from aluminum siding, with a makeshift wooden awning above the front door.

In the lot to the left of Trek is a massive airplane hangar, like something Howard Hughes might've owned, which is home to a company called California Sport Aviation. In front of an open sliding door that's taller than a pair of elephants, about seventy yards away, two guys tinker with a red-and-white single-person light airplane roughly as big as the red Saturn belonging to Trek's Rob Bulaga, parked in front of Hangar 21.

A sound like that of a tractor engine pierces the air. "Jesus," says Dad, giggling. "I think he's going to take the damn thing off right here." Neither of us has spent much time around small aircraft (despite the fact that Dad's father was a radar operator for the air force in World War II), and it seems slightly odd that a person can simply get into the air, just like that, with very little fanfare and no stewardesses. Dad cracks, "I get the eerie sense of someone about to die, er, fly."

Just then Rob Bulaga emerges from inside Hangar 21 and ushers us inside. We miss the takeoff. To be honest, Rob seems a little depressed by the current state of Trek's HQ—the tight quarters are thick with dust and tools, gadgets, electronic parts with their guts exposed, bolts, nuts, and cranks spread across every empty surface. I'm not sure why, though, since this place is pretty much an adolescent boy's fantasy come to life and he gets to work here. Maybe he's just tired; it is still before ten o'clock, after all. Rob has darkly circled eyes and thin, graying hair worn in a sharp part. He has on the blue-collar engineer's uniform, a thick, sensible button-down flannel shirt tucked into light-blue jeans, and white sneakers. A pocket watch dangles from a loop on his belt-less pants.

After studying engineering at the University of Illinois, Rob took his wife and two daughters with him to St. Louis for a job at McDonnell Douglas. A few years later, the family headed west, and he landed at Trek, back when it was based in a sprawling office park in Sunnyvale.

Now Rob is pointing at a college dorm refrigerator in one corner and the small, spent coffeepot on top of it. "That's the cafeteria," he deadpans.

On a dry erase board someone has drawn a mechanical something that looks part shark, part helicopter. Under the doodle it reads, "There is no cure for stupid." In another corner, the side of a bright-yellow box the size of a television set from the '80s warns, "Flammable Keep Fire Away."

Rob is the only one of the three main Trek staffers at work. Company president Harry Falk, who handles finances and promotion, and Tim Worley, who handles everything else, are not around. It's quiet inside Hangar 21.

We walk over to a Springtail prototype, next to a standing Craftsman's toolbox that's easily big enough to double as a phone booth. "The controls of the Springtail are like a joystick—it's like playing a video game," Rob says, pointing to the motorcycle-like hand controls that remind me of Bell's Rocket Belt design. "The left one is the throttle; the right is for yaw and control. When you pull your hand up, it feels like you are levitating."

Dad looks at me, buries his top lip under his bottom one, and bobs his head rhythmically as if to say, "I like the sound of that!"

The Springtail is eight feet tall, nearly nine feet wide, weighs about 370 pounds, and can hold 10.6 gallons of gas. Things were looking good for Trek a few years ago—they were making progress, getting great press, and living high off the DARPA money. But in December 2002, while testing the Springtail between storms, the company took a big hit. Rainwater seeped into a bungee cord connected to the steel tether line, causing the bungee to stretch and slacken. Now the tether had too much slack, and as Rob cracked the throttle and rose off the ground four feet, the loosened wire bunched up near one of the Springtail's massive duct fans. Blade met steel; steel lost. Rob tumbled straight down but was mostly unharmed. The same could not be said of the Springtail, which suffered about eighty thousand dollars in damages. DARPA froze Trek's funds. "A car parked eighty feet away had its antenna dented by debris," Rob says. Dad and I stare at Rob solemnly, marveling at the bad luck of it all.

It was soon afterward that Mike Moshier left the company. Since then, even good news has had a sour aftertaste. When producers for the film *Agent Cody Banks,* starring Frankie Muniz as a teenage CIA officer, asked to use a SoloTrek in the production, Trek happily complied. The only money from the deal would come from future action-figure sales, but, in the end, the figures were never made. The movie's poster is, nonetheless, prominently displayed near the hangar's front door. Stars Muniz and Hilary Duff make too-cute faces in the foreground, while behind them in the middle distance a tiny, glowing SoloTrek hovers. "Muniz is fighting the bad guys," Rob is saying, describing the SoloTrek scene. "He tricks him and straps him into the SoloTrek, pushes a button, and sends him to the CIA headquarters."

"So that model can fly autonomously?" Dad asks.

"In the movies, yes," says Rob. We all laugh uneasily.

Next to the *Agent Cody Banks* picture hangs a Trek promotional poster. On it, Rob is strapped to the Springtail, wearing a crash helmet adorned with a microphone headset, sunglasses, and a look of intensity. Behind the machine, looking similarly intense as he gazes up at the pilot, is rail-thin Tim Worley, the company's electronics expert. Tim is also wearing a helmet and clear protective goggles. He has on dark jeans and a gray Mickey Mouse sweatshirt. The ends of his thick ponytail ride up on his left shoulder. Tim has heavy work-men's gloves on—and his hands are held apart, knees flexed. He is hunched, ready. The Springtail is hovering about half a foot off the ground, but my eye keeps coming back to Mickey Mouse's famous profile. And then I see Tim again—intense, prepared for anything. I begin to think that Tim seems very capable and am filled with new hope for Trek.

Then we sit down at Rob's soot-coated desk, one of four desks packed into a corner of the hangar, to watch some video on his computer. From the screen comes a loud whir and the sight of Rob elevating a few feet off barren roadway. "If I don't double-knot my shoes, they come untied," he says, rhapsodically.

A PowerPoint presentation begins, featuring images of Trek in the pages of *Time* magazine, in Ripley's Believe It or Not's *World's Weirdest Gadgets* book, and then a shot of the Springtail's first untethered flight on December 18, 2002. "Ninety-nine years and a day after the flight of the Wright brothers," Rob notes. "Ah, well, Mike always said, 'If it was easy, someone else would've done it already.'" Outside, a plane from the California Sport Aviation hangar roars by so close it sounds like it might take Trek's doorknob with it. "If we can survive the next year, there's a good chance that we can get something on the commercial market in five years," Rob says. He notes that there is a large sheriffs' convention coming up, and he hopes to makes some OVIWUN sales there. The idea is to affix a small camera or flashlight to the battery-powered device. The OVIWUN's duct fans, about the same diameter as a large grapefruit, would then lift the drone into small city crevices and around creepy corners of sketchy urban areas so that officers would not have to tread dangerously. That's the idea, anyway. Rob sums up the company's financial outlook: "We're still living off the DARPA money. But it's running out."

I look around the area by the desks. There are four wall calendars, all open to different months, and two of them were made for other years. On top of a large filing cabinet, a model airplane sits under a web of dust. Hanging over the front door, held by metal cables, is Trek's final fantasy—the Dragonfly. It's like the cockpit of a fighter jet, detached from the rest of the plane, and tricked out with those same duct fans. It weighs 490 pounds without any cargo and is designed to cruise at about 170 mph at up to 12,900 feet in the sky for three hours. This prototype was thrown together in four months and rushed to the annual special-forces equipment show in Tampa in 2004. It looks not to have budged much from its dangling perch since then.

I like Rob Bulaga, I really do. Something about his kind face and sad eyes and his slight paunch makes me want to root for the guy. But standing in that tiny hangar right now, I just don't feel like I'm tiptoeing toward the edge of the future. Those two calendars on the wall are not the only things stuck in the past.

We say good-bye. I wish Rob luck with the upcoming sheriffs' convention. He walks us out, takes a quick peek at the looming hangar belonging to the California Sport Aviation people, and disappears back inside.

Dad and I are due at Moller's Davis headquarters at 1:30 for a brief chat with the company's general manager, Bruce Calkins, before the weekly open-to-the-public tour of the facility. That gives us enough time to get into town and grab lunch before checking out a flying car. And it's not just any lunch—today's Dad's sixtieth birthday, so we should celebrate.

An hour later, we are cruising through downtown Davis. This is the real California—vast blue sky, squint-forcing sunshine, block after endless block of fast food shacks and gas stations. I want to go someplace nice for Dad's big birthday lunch, but I know he doesn't care about such things. I try to push for the better of two burger franchises, but he'd just as soon go to the one on the way, which means we'll have his sixtieth birthday lunch in a Wendy's across the main road from the UC-Davis campus.

Dad never looks right when placed in the most banal buildings and environments of modern suburban sprawl. He'd look much more at home in a library or lab, or on safari. In Wendy's, with his standard black-on-black ensemble, his dark, mirrored shades hiding his eyes, and the silvery puff of his hair, he looks like a benevolent space alien.

He considers the menu as if it's the first time he's done such a thing. After I explain that a Frosty is just Wendyspeak for a milkshake, he tells me to order one of those for him. But then he changes his mind and asks for the same as me—cheeseburger, fries, diet Coke—then scampers off to find a table by the window, the better to muse upon the "dance of the metal dwarves," as he once referred to the traffic out the window.

By all rights, this should be depressing. Dad is sixty today, and we are eating bad burgers in a strip mall next to a BP gas station. So why am I sort of elated? I realize it's because Dad has a way of making everything an adventure. Even here, where every last fry is scripted, I still have the sense that anything could happen. It's been like this as long as I can remember. Even before Mom moved out for a year to get her teaching degree in New Hampshire, leaving us alone with Dad, I never could predict what might happen next. Some would say this is probably not the best way to raise a kid, but I remember being mostly thrilled by the arrangement.

A typical Dad adventure was his "tour of Baltimore." I was about eleven and thought, "A tour of Baltimore? A tour? Of the place I've lived my whole life? What could this mean?" I couldn't wait for the weekend. Saturday finally came, and my brother, my dad, and I piled into Dad's Volkswagen Beetle, the color of overripe bananas. During the school week, the car felt like a travesty in the pickup line. It practically screamed "scholarship" to the other moms and dads in their black Beemers, Mercedes, and Cadillacs inching through the carpool lane of our private school's campus, a woodsy, bucolic paradise.

But on the weekend, I secretly loved the clown car with the sun-roof, which we opened on the brilliant, crisp fall afternoon of our "tour." Dad showed us his favorite street—a curvy downtown stretch running beside a rusty slalom of train tracks. And his favorite view, an overgrown baseball field with a clearing, high above the silent city. And his favorite drinking spot, tucked in among the old town's cobblestone streets with a long, shiny wooden bar, plenty of droopy ferns, and zaftig, doting women.

Now, here in a downtown Davis Wendy's on his sixtieth birthday, I could feel the same giddy uncertainty bubbling inside me while I watched him devour his burger, mayo glistening on his lips. We read sections of the *Sacramento Bee*, commenting now and then on how insane the world can be.

Then we are back in the truck, heading across the street to the block-long wooden ranch-style spread that is home to Moller International. It's part of a research park that Paul developed in the early '80s, kick-starting his real estate business. Across a meandering campus roadway is the neuroscience building.

If a forest gnome opened an orthodontic practice, it would prob-ably look a lot like Moller's digs, all sleepy '70s architecture, a nar-row, winding sidewalk leading to a glass front door. Though it is still two weeks until Halloween, the lobby is ready; employees are psy-ched. A fake spider web covers the receptionist's desk, purple, or-ange, and black plastic spiders spinning within it. Frankenstein wishes us a "Happy Halloween" from a life-size poster, adorned with long tassels hanging across the doorway leading to a conference room. On one wall, a four-by-four computer-rendered poster effec-tively has a Skycar zoom above a futuristic field of wheat. On the op-posite wall is a staged, stylized photograph of a cop ticketing a parked, passengerless Moller creation on Main Street, USA. It's not clear if the Skycar's driver is in a no-parking zone or is simply get-ting a ticket for being too cool.

A few minutes pass before Bruce emerges from the back of the office. His walrus-y mustache and amazingly thick head of hair remind me of oatmeal and comfortable sweaters. I didn't realize it until now, but I may have been a bit nervous about this encounter—Bruce sounded brusque on the phone, in telling me that Paul was too busy to meet with us. But now, seeing him, I am put at ease—you can't have a mustache of that caliber and be a bad guy.

We push aside the Frankenstein tassels and enter a typical conference room. Well, it's typical except for the fact that the long table's centerpiece is a sixty-pound, 120-horsepower, four-stroke rotary engine.

"Has that been modeled or is that empiric?" Dad asks, pointing to the centerpiece.

"Um, yes," Bruce says. They explode in laughter while I try to catch what was funny. We're in.

Bruce launches into the history of Paul. "The story we tell is that he was walking through the snow," Bruce began in his Wilford Brimley baritone, stopping for a second to play an invisible miniviolin, a would-be somber soundtrack for the melodrama at hand. "He saw a hummingbird and said, 'I want to be a hummingbird.' As a child he was very mechanical—and living in rural British Columbia on a chicken ranch, there were always lots of [car] parts around, so he figured he could blow things up." At fourteen, he built a working car out of discarded parts. Before he turned sixteen, Moller was renting and flying a helicopter for fun. According to the *Los Angeles Times,* when he was nineteen he came across the archived specs for a machine that was jointly produced by the British and Canadian governments in the 1950s. The Avrocar was "a flying-saucer-like vehicle designed to use the thrust from a horizontal, center-mounted propeller to zip into the sky." The Avrocar was meant to be a "nimble military air carrier that could shuttle troops," with a civilian spin-off dubbed the Avrowagon. Neither vehicle could be made to work. This sounds terribly familiar.

Moller followed a trade-school diploma in aircraft maintenance with a second diploma in engineering from a school in Calgary, Alberta. He then obtained a master's degree and Ph.D. from McGill University in Montreal in mechanical and aeronautical engineering. There he met his first wife, a psychology student named Jeanne LaTorre. They married on New Year's Eve of the same year and soon after packed up their lives and moved to Davis, California. It was 1963—the university was intrigued by his ideas for a flying car.

And that's when Paul Moller built in his garage what in photos looks like a classic late-night-movie UFO out of fiberglass and aluminum. The XM-2 could not do more than hover at six feet in the air, but it made Moller locally famous, and it provided the technological foundation for the work he has done since.

In the '70s, Moller began experimenting with duct fans, strapping them to bikes, skis, people, you name it. To combat the noise, he developed a muffler, which, as fate and fortune would have it, led to the hugely successful SuperTrapp muffler company. When he sold part of that business, he used the cash to start working on the Skycar. Today, he owns 64 percent of the publicly traded company that has about 650 investors.

"The Skycar is a story of engines more than anything else," Bruce is now saying. This is all the prompting Dad needs, and he asks Bruce if he can pick up the rotary job in front of us. Bruce says he's fine with that. Dad waddles behind Bruce and around to the lip of the table to get a decent purchase on the sixty-pound brick of power. He lifts it, with a grunt, out of the mount. "Oh yeah, oh yeah," Dad practically moans, shaking the engine slightly up and down, as if it's a melon whose freshness he'd like to determine. Bruce looks a little anxious.

Dad holds the thing for what seems like an hour but is probably only two minutes. I keep glancing at Bruce to see how he's taking it. He's smiling but his eyes say, "Please just put it back. Please." Finally, Dad does—or at least tries to. But the thing won't go back in the mount, and he's struggling to keep it from crashing through the faux-oak conference table. Bruce's hands shoot up reflexively to help, but he stays seated.

Then Dad has an idea. He picks up the mount off the table and fits its long steel pole into the engine socket, so the mount is balanced on top. Then he ever so gently rotates the entire mass, flipping the mount to the underside of the engine, and lowers it all down as one. Bruce relaxes back into his seat. Dad's brow is a little moist as he returns to his.

Bruce wraps up his oral history: "Survival is always an issue for a small company, and we're almost the epitome of what it means to be small and disabled. But despite what people think, we haven't been sitting around for thirty or forty years—we've been progressing as best we can without killing anyone."

This doesn't strike me as the impassioned rhetoric of the company of tomorrow. Don't we need to hear more about American drive and determination, not resting until the job is done and done right? All that stuff?

"To get the Skycar in the air," Bruce concludes, "we anticipate we're going to need about two hundred million dollars, but I fear it's going to be more than that."

Oh, man—let's just start the tour, shall we?

Dad and I follow Bruce back out into the small carpeted lobby, where a handful of youngsters and their parents are assembling for the afternoon spin around Moller International. Padding down the hall, the group emerges into a larger back room that was built in the '80s but feels like the 1950s. Maybe it's the model of Moller's flying saucer in a display case. Or the multiple Skycar models under glass. Or the black-and-white photos of Paul looking like an Amish spaceman actually flying the thing. Or maybe it's that line of old putty-colored filing cabinets, labeled "Aerobatics, Inc.," "Patents," "Freedom Motors."

In any case, now that there's more room for the tour, I can take stock of the type of people still believing, or wanting to believe, in the forty-year dream. There are fifteen of us, including Dad and me. There are seven wholesome middle schoolers with buzz cuts and looks of wonder, here with three sensibly dressed women who are either mothers or chaperones or both. One middle schooler is a dyed-in-the-wool Super Geek—polo shirt buttoned to the top, braces, pipe-cleaner thin, carrying a clipboard. A Pakistani couple has brought their two boys, one of whom is a one year old in a Spider Man basketball jersey and matching shorts. A single tween girl wanders with a cluster of young dudes.

Bruce shuffles over to a display case. A corn-fed kid asks about flying the saucer.

"Ever tried to stand on a basketball?" Bruce asks rhetorically.

"A few times," Super Geek says, not smiling. I can't tell if he's joking, and no one laughs.

The Pakistani dad says he heard the company is taking deposits on Skycars, to be delivered on December 31, 2008.

Bruce sighs. "Well, that's what it said on a document we put out a couple of years ago." A group guffaw.

Moving over to another case, Bruce shifts gears. "Another thing we have here is a flying robot."

The lone middle school girl looks up. "Nice."

Bruce leads us into the main engineering room. I straggle behind to admire Moller's wall of press: a mention in *Esquire*'s 2003 "Genius Issue," in an article titled "America's Best and Brightest," the *Los Angeles Times* magazine cover story introducing readers to "The Audacious Mr. Moller," a *Popular Mechanics* cover from 1991, and a *Popular Science* from 2000. There are many others, all crowding a pool-table-long corkboard. Paul's PR Department has certainly done its job.

I follow the tour into a cavernous room in the middle of which is what could be a giant metal worm, escaped from the set of *Dune* long ago. It is actually a wind tunnel, and even though, architecturally, it isn't much to look at, I gasp. I peer through one side of it. It's long, maybe forty yards, and quite narrow. More like a wind tube than a wind tunnel.

We walk around the tunnel, and there it is—the M400 Skycar. It's the latest iteration of a four-decade obsession. I have to say, the thing is really, really attractive. The press often compares it to a Batmobile, and I can see why. It's got to be the closest I've ever been to a World's Fair future. The M400 is straight from the deco wonderland that sprang fully formed from the imagination of artist Arthur Radebaugh as far back as the 1930s. Gorgeously sleek, hypermodern, impossibly cool. If I can't get a jetpack, I think, I'll settle for a Skycar.

But now Bruce is losing the group. He's talking about how the next model will have new high-powered computers installed for better, more reliable control and how beefed-up technology will allow for about 450 pounds of thrust per engine. The buzz-cut bunch is shifting impatiently on their heels. A young guy in a San Francisco Giants cap pantomimes his pickoff move. The middle school girl is sitting on the stone floor, looking at her shoelaces. Only Super Geek and my dad are still with it. Dad's walking the length of the Skycar, admiring its shape, stroking it here and there.

"When are you going to build the next model?" Super Geek asks Bruce, meaning the one that will fly, pending further tests, FAA red tape, and that two hundred million in production and certification costs.

Without missing a beat, Bruce fires back, "As soon as you give me the money." The moms and dads chuckle; Bruce has clearly heard that one before.

Super Geek (sheepishly): "Okkkkaaaayyyy."

An old man shuffles past our tour, slumped shoulders, haggard, paunch like a cantaloupe. Paul Moller—the genius. He doesn't stop to say hello to the kids, and Bruce doesn't bother to introduce him. Moller pokes his head into an adjacent workroom, calls out a name, and when there is no answer, he leaves, a defeated Willy Wonka.

Trying to reclaim his charges, Bruce mentions the new control system, how it will be like playing a video game. "It will be so easy for this generation to fly a Skycar," he says, nodding at the fidgeting blue-eyed army. "Because they will be acclimated to joysticks and mouse clicks."

And just like that, against all better judgment, I am a believer again. It could happen. We could all one day take to the sky. Well, maybe not me, but surely my kids.

Our half hour is up, and the crowd starts heading back into the first main room to buy Skycar models and T-shirts. I corner Super Geek and his mom, a lively-eyed Davis lady. I mention how they seem really into this stuff and ask if they have any thoughts about jetpacks. "I'm forty-five," Mom says. "And I remember *The Jetsons* and *Back to the Future,* and their future was 2001. Well, we don't have it, do we?"

No, we sure don't. Super Geek doesn't have much to say about jetpacks, but he does tell me that he's been designing a flying car himself. Thus the clipboard. "It's kind of like this." He flicks a hand at the Skycar. "But my design is more like the one in *Back to the Future,* only with wheels that can fold up."

I tell him that sounds cool. Really cool. And to put me on his mailing list. He says he doesn't have a mailing list.

Back in the room with flying saucers under glass, I buy Dad a Skycar model the size of a matchbox for his birthday. We say good-bye to Bruce and head back to the white pickup.

We make good time back to Walnut Creek, both of us thinking more than talking. I look out the window and watch as loamy farmland blurs into the suburban film loop of Target, Starbucks, Barnes and Noble, the next Target. The day has been bittersweet for me. Although it is encouraging and inspiring to meet true believers of the air, the truth is that neither the Springtail nor the Skycar will be a viable jet-pack alternative anytime soon. The search continues.

I turn to Dad and ask what he thinks. I'm hoping he can make sense of what we've seen today and help put my quest in perspective. He knows his history, his physics, and his Wright brothers, after all. Given this, does Dad think any of us will really ever fly to work? I ask him if he wouldn't mind jotting down a few notes on the subject. His answer, as always, surprises and intrigues me.*

At the moment, however, we are at the Walnut Creek BART station and he is pulling the white pickup over to the curb to drop me off. My head is swirling from our day out at SoloTrek's HQ and to see Paul Moller's not-quite-flying car. But I must focus: Catherine, the girls and I are moving to my brother in law Joey's family's house in San Fran-cisco tonight and I'm heading into town early to get ready for Jimmy's North Beach bachelor party. Dad and I make hasty plans to talk be-fore my Mexico trip, four days away. He lets the car idle and gets out to say goodbye. Once more I am in his big, pillowy embrace, the sweet-sour smell of his skin billowing up all around me, and it feels like a sucker punch to the heart: I'm surprised at how much I know I'm going to miss him.

The San Francisco days pass in a haze made of equal parts rolling celebrations and serious sleep deprivation. Once again Catherine, the girls and I are sharing pretty much just a single room, this time in a basement apartment underneath Joey and Anne's three-story house, teetering on one of those famous hills not far from the Mission Dis-trict. And so, bleary-eyed, I stumble through a socializing Ironman competition: the obligatory too much excessive booze and sketchy strip club of the bachelor party, a lovely wedding (from what I saw of

---

*To read what is truly an engaging and fascinating document on the physics of flight (yeah, okay, he's my dad, but still, I think you'll enjoy it), check out www.jetpackdreams.com for the Web-only bonus material.

it) high in the East Bay hills, spent chasing Oona and bouncing Daphne, and the post-wedding backyard BBQ.

When we moved to Brooklyn from Portland, Oregon, Catherine and I agreed we'd stay for five years and then trek back West to the Bay Area, where several of her siblings were already beginning to procreate. So many cousins for our girls. We've been in New York seven years now and it's getting harder and harder to rationalize keeping our two kids away from so much family. The few days in San Francisco have only reinforced the idea that the clock is ticking. This is just another Big Life Decision I need to file away for now because tomorrow I'm leaving for Cuernavaca, Mexico, and my meeting with rocketbelt legend Juan Lozano.

The plan is that I'll fly from S.F. to L.A., connect with my old Baltimore buddy, Jeremy, and then the two of us will continue on together. We'll stay overnight at Juan's place and then one night in Mexico City before flying back to Los Angeles where I will meet up with Catherine and the girls at Nancy and Rob's rehearsal dinner party in Santa Monica, the night before their wedding. Whew! The morning comes too early. In the pre-dawn moon glow, I hold my breath and ease off the bed, where Oona and Catherine are both asleep, cheeks mashed against pillows. I shower, dress, and pack, then lie down again for a minute before leaving.

It's dark in the room but I can hear Oona's deep nose-breathing and can smell the sweaty sweetness of her scalp. Then her voice startles me:"Do tickle face, Daddy," she whispers through her dreamy fog.Tickle face? I know of tickle face but as far as I can remember had never actually been responsible for providing tickle face. Until that moment, in fact, I'd been pretty sure that tickle face was Catherine's department, while I handled such things as Silly Walks and Cozy Dances. Tickle face? I'd give it a try. Gently, like a blind person, with the tips of two fingers I trace Oona's forehead, cheeks and ears in the coalmine blackness. I make tiny zigzags over her chin. I have no idea if what I'm doing is authentic tickle face or when enough is enough—how does one know when to stop tickling the face? How long is too long?—I sure don't want to wake her. This is a rare moment when Oona and the Byrd are both asleep. I could actually do this forever, I think—it's the most tender, innocent thing, as are her little-kid snotty snores. So I tickle her face.

But I also don't want to miss my flight. I hold my breath and, again, ease off the bed. I grope around near the floor until I find the handle of my suitcase, turn to tip-toe out of the room—and step on a plastic barn, which moos incredibly loudly in the dark. Shit! I freeze. Listen, holding my breath. There's a slight stirring; someone turns over. Then, quiet. I avoid any other farm animals on my way down to the street and out into the chilly San Francisco air. It is still very early—before sun, fog, squirrels. So quiet. I roll my suitcase down the hill toward BART, Los Angeles, Jeremy, Mexico, Juan and, with any luck, my first-ever wingless flight.

CHAPTER
5

South of the Border, North of the Future

You can fly! You can fly! You can fly! You can fly! Soon you'll zoom all around the room. All it takes is faith and trust. But the thing that's a positive must is a little bit of pixie dust. The dust is a positive must.

— **Peter Pan**

"Maccabee Blake Montandon, as I live and breathe!"

"Jeremy, how are you?

"Good, man, good. I've been practicing my Spanish so when we're in Guadalajara in the back of the truck with the chickens, we'll know what to do."

It's a few minutes past nine in the morning at LAX, and Jeremy has just awoken from a nap, which he desperately needed after pulling an all-nighter working as a freelance film editor to support the horror films he wants to make.

I've known Jer since ninth grade when he transferred to my small, progressive private high school from a theater academy. He's always avoided the pedestrian—in his taste, look, attitudes. Back then that meant '50s ties, sharkskin pants, pointy patent-leather shoes, and a Mrs. Robinson affair with a family friend. Today, though he's traded the sharkskin for jeans, he's kept the devilish pencil mustache, soul-patch combo, and shoulder-length, tightly curled black hair.

We became fast friends, and I spent the better part of subsequent Baltimore summers holed up in the attic bedroom of his suburban house, drinking, smoking, making out with girls, and trying not to wake his parents.

Along with his younger sister, Jer and his folks did well for them-selves in the local dinner theater scene. On the living room walls hung several commemorative posters from some of their better work, *Carousel, Oliver!* and *The Music Man*, the four Kastens smiling mas-sively, heavy makeup turning cheeks to dimpled radishes.

Jeremy transferred midway through our sophomore year after get-ting busted for smoking pot in the woods. He went back to art school, then studied film at a Boston college before bolting for Hollywood. He's lived there the past fifteen years and refused to give up on his directo-rial dreams. With the help of a small team of producer friends, Jeremy's already raised funds for, shot, and completed three minimasterpieces of the B-movie horror genre: the disturbing psychodrama *The Attic Expe-ditions,* starring Seth Green; an *Evil Dead* homage, *All Soul's Day;* and *The Thirst,* featuring that crazy guy from *Six Feet Under.*

As we board a Boeing 737 for Mexico City this morning, Jer's in post-production on his latest scare fest, a remake of the kitschy 1950s blood-bath called *The Wizard of Gore,* starring the professionally bizarre Crispin Glover. The film will also mark the big-screen coming-out party for a collection of online Goth-punk soft-core-porn nymphets known as the Suicide Girls. My old friend has thus been spending a lot of time offset hanging out with a small but busty group of Suicide Girls, which makes me think that perhaps there really is such a thing as destiny.

The last time he and I took a road trip together was during the sum-mer after our junior year in college. We were both out west, visiting my brother, who'd moved in with Dad in Oakland and joined a Rocky Horror live troupe, singing and acting out the midnight-movie cult favorite in real time. The perpetually stoned performance group had been hired to do a show in Monterey. Jer and I thought it would be a good idea to drive down with a gigantic and hairy *Rocky* hanger-on, who also happened to be a former psychiatric patient of my Dad's, in part because he genuinely believed he was the cartoon character Captain Caveman, but mostly because of his fondness for crystal meth. The Captain, as it happens, hadn't slept in a week and a half and ended up wigging out, accusing us of stealing his pot, and then parking his van on a very desolate stretch of California roadway, telling us he wasn't budging until we confessed. "I have no problem dragging both of you to the back and bludgeoning you to death!" the Captain bellowed as the hours dragged on.

Problem was, we had nothing to confess. I grew fairly certain that I had seen my family for the last time. There appeared to be no way out. The rusty brown-orange van we drove could have been lifted off the set of any number of horror films from the 1970s. Finally, out of fear and boredom and desperation, I broke down and told the Captain I did it, fully expecting the bludgeoning to commence. "OK, that's what I thought. I'm glad you finally had the guts to say it. Now, should we join the others for dinner?"

Was he fucking kidding?

So now Jer and I are back in the saddle—yeehaw! As I suggested earlier, this is why I consider him my ideal travel partner and camera crew. We have a history of survival. On the flight down, we munch pulled-pork sandwiches and toss back a couple of plastic cupfuls of cabernet sauvignon. We talk about movies, magazines, the past. When I bring up his theater days, Jer reminds me that he once played the boy, John Darling, in a substantial and legit downtown production of *Peter Pan.* "Dude, perhaps my greatest thrill of all time was being eleven years old and knowing every night I would get to fly! How fantastic is that?"

Jeremy is not the most outwardly emotional person I know, yet he appears to be tearing up. Thoughts of flying can do that to a guy, I guess. Or maybe he's just exhausted. Before I know it, I've dozed off myself, and then we are there.

As we enter the Mexico City airport Jeremy announces, "You realize, of course, that I'm going to add a bit of Hunter Thompson to the proceedings?" I don't doubt that.

To begin with, this means hunkering down at the airport's Freedom Bar for a shot of whiskey before phoning Juan to let him know we've arrived. I can't help but think that the bar's piped-in music is trying to send me a maudlin yet meaningful message—consider this trio of hits strung together: a Muzak version of *Dream, Dream, Dream,* then Rod Stewart's *Forever Young,* followed by the '80s reggae-dipped Culture Club ballad *Do You Really Want to Hurt Me?*

Speaking of pain, a few days before we were due in Cuernavaca, Juan e-mailed me, cryptically saying there had been an "accident," that he was "just out of the hospital," and thus wouldn't be able to pick us up at the airport. We were to take a bus an hour and forty minutes south, then call from the Cuernavaca terminal, at which point Juan's wife would pick us up. He hadn't elaborated, so I'd feared the worst, that is, a rocket-belt crash landing, dashing all of Juan's dreams, maybe even permanently damaging his body and mind, not to mention all of my hopes for a first flight.

Jer and I down our whiskeys and head for the door. The bus creeps away from the airport, lurching through Mexico City's notoriously bad traffic. But soon we are slithering across the Sierra Madre, which could be a family of enormous, sleeping purple dinosaurs in the fading daylight. By the time we come down the other side and begin descending into Cuernavaca, night has arrived. The town is popular with retired Americans, which makes perfect sense as we pass a sign proclaiming, "Cuernavaca: Cuidad de primavera eternal"—City of Eternal Spring.

The bus hums along broad boulevards, busy on both sides with short-sleeved locals crowding leafy esplanades, pouring out of Technicolor-bright apartment buildings, Taqueria Leo, Wal-Mart, Bodega Gigante, the La Roca nightclub, and, finally, a car dealership, still bustling at seven thirty at night.

We phone Juan as planned, and twenty minutes later his wife, Isabel, arrives in a silver minivan with a TAM sticker on the back: Tecnologia Aeroespacial Mexicana, the company Juan has set up to sell hydrogen peroxide distilling machines, his rocket expertise, and even, for a cool $350,000, an actual working rocket belt.

Isabel and Juan met when they where teenagers growing up in an upscale neighborhood in Mexico City. It was hardly love at first sight. "He was always with his mother," Isabel says in the van, scrunching up her nose. "Ech." We turn on to a narrow, darkened street. As time passed, however, Isabel felt her heart softening, and then one day Juan invited her over to watch him test his very first backyard rocket. A fuel malfunction caused the rocket to explode at takeoff—flying debris whizzed past Isabel's face. Love bloomed. They were married a few years later, quickly had two daughters, and moved to the quieter, less smoggy Cuernavaca about twenty-five years ago.

Juan went into the family jewelry business, eventually earning enough pesos to pay for the sprawling, gated compound we are headed to now. Besides the two-story tangerine-colored house, the property contains four studio rooms ringing the verdant yard. Visiting students at the local language school stay with the Lozanos year-round; Isabel cooks for them and shuttles them back and forth to class.

Juan's girls are grown now. Isabel, named for her mother, is an architect and married with a two-year-old son. Claudia is in music school, where she plays piano, guitar, bass, French horn, and trumpet in the band.

"And here we are," says Juan's wife, turning onto their street. She is short and slight, with dark mascara under her eyes, a broad, frequently deployed smile, and a face wrinkled, so it seems, by years of laughter.

Trash lines one side of an otherwise well-kept suburban road—the result of a strike, Isabel explains. "Eet ees 'orrible," she says, an expression I'll hear more than once that weekend.

Isabel clicks a button, and the garage door opens. She parks and leads us into the shadowy backyard. I can make out the darkened shape of a long-branched avocado tree. We turn at the tree and begin climbing a short flight of half-finished concrete steps. I'm still not sure if Isabel is taking us to our room or to see Juan. At the top of the steps, Isabel opens a door, and we enter an unlit room. My eyes are adjusting, beginning to make out the blackened, lumpy shapes around us, when I hear a deep voice ask, "Hello, how are you?"

Juan Lozano, the Mexican Rocket Man, is sitting in a La-Z-Boy chair in the near corner of the room, his feet propped on a cushion. He flicks on a floor lamp, and I see that he is wearing a gray T-shirt pulled taut over a big belly, short black shorts with the words *Racing Team* written over a checkered flag, and checked slippers. He has Ace bandages wrapped around both thighs and his right calf. His arms are badly scabbed. He is practically bald, a nimbus of whitening hair at the back of his head, and has dark circles under his eyes. I've come to meet the rocket-belt Buddha, to pay my respects, and now I feel as if I've stumbled upon Cuernavaca's Kurtz.

Isabel excuses herself to make Jeremy and me dinner. Mexican soft jazz tinkles from a cheap portable cassette player balanced on a bookshelf.

"Juan," I say, "what happened?"

He gestures painfully for us to pull plastic chairs from a nearby desk and sit. Then Juan begins his story. It was the weekend of the convention in Buffalo. Rather than deal with the hassle of getting one of his rocket belts through customs, Juan decided to take his wife and a rocket bike he'd built to a demonstration at an engineering conference in Acapulco. A half hour before they were to load the bike into the silver TAM minivan, Juan thought he should test the controls one last time. So at about nine in the morning, he gathered a handful of friends and relatives and pushed the bike with an 1,100-horsepower rocket welded to it onto the street in front of the home he'd designed to give his family peace and comfort. Juan knew the rocket bike could really fly, but he was curious to find out just how fast. This was to be a quick test, however, so he hadn't bothered to put on a racing suit—just a helmet, shorts, and a T-shirt. He cranked open the throttle. The bike exploded down the block like a bullet from a Saturday night special. Neighbors later said they thought a bomb had detonated. Juan thinks he hit 75 mph in a little more than a second. He's not entirely sure, however, because things got a little foggy after that.

The back wheel couldn't take the power, the pressure, the speed, and crumpled, sending Juan skidding hard along the pavement. He was out cold—and Isabel was beside herself with fear. Juan was whisked by ambulance to the hospital and put back together like a lucky yet very unbionic man with four right ribs broken, a snapped clavicle, and so much skin scraped off his thighs he had to have calf flesh removed and grafted back to his upper leg. "I told the doctor, 'Take dog skin, take pigskin—anything!'" he tells us, his eyes widening finally beneath heavy lids. Then, his voice softens. "The doctor said I could have been a paraplegic." His English is good, if a little lethargic due to the fistfuls of pain pills he's been guzzling for a month. He picks up a motorcycle helmet near the chair—it's cracked clean down one side, as if he'd been attacked by a baseball-bat-wielding psychopath.

I realize I've been literally on the edge of my seat since Juan first began his story. I sit back now and come to terms with a few things. One, Juan is damn lucky to be alive; two, he doesn't seem that shaken up—perhaps enough time has passed for him to calm down; three, I may not get that first flight after all. It's only then that I look to my left and notice the four perfectly sleek rocket belts mounted on stands just a few feet away. Even in the shadows, I can make out enough exquisite chrome, enough space-age craftsmanship, that a static shock of excitement jolts the hair on my arms. Beyond the belts is a large putty-colored contraption as tall and wide as an NBA power forward. It contains a round glass science-class beaker and much tubing connected to what looks like a refrigerator from an imagined future. Or maybe the 1950s.

Juan yawns a massive, slow-motion gape that wrinkles up his smooth, mocha-hued facial features. He looks at us, we unburned, un-broken gringos, with an expression of, "So, what can I do for you now?"

Juan Manuel Lozano was born in Mexico City in 1954. He was one of three children, and the only son, born to a successful jewelry designer and retailer and a doting stay-at-home mother. When little Juan was seven years old, his mom took him to the Campo Marte polo grounds to see demonstrations as part of NASA's American space expo. The expo included a Mercury-capsule exhibit, astronaut gear, and samples of chalky space food. But all Juan wanted to see were the twenty-one-second Bell Rocket Belt flights. And he wanted to see them over and over and over again.

Bill Suitor and Peter Kedzierski alternated liftoffs during the week-end expo, blasting the belt into the atmosphere four times a day. With each flight, the machine captivated little Juan Manuel a bit more.

He went back to the campgrounds the next day, his mother pack-ing homemade peanut butter sandwiches to maximize the time her son could spend with his first full-blown crush. And Juan didn't squan-der his chance to see the belt. For some flights he'd stand on the same side of the field as the pilots; for others he'd move closer to where they were landing. When the pilots rested, Juan did not. He spent the min-utes between demos inching as near to the science fiction fantasy come to life as he could. As the machine rested on its stand, the boy traced each part with his eyes, the glistening fuel tanks, the narrow nozzles, the otherworldly throttle, the jetovators. "I don't remember what I did last week," Juan once told *Popular Science.* "But I will never forget the first time I saw that Rocket Belt." How could he? He's spent more than thirty years and half a million bucks attempting to fly one. But that's getting a little ahead of the story.

When he was in high school, Juan would often skip class to go work in a local race-car manufacturer's shop. "The most important things I learned, I learned there and not in school," he tells me. "I learned how to weld, how to make the molds, how to work with fiberglass, to design, to convert engines. Everything."

All of these skills came into play years later when Juan finally had time to get serious about building a belt. He went to school for his commercial airline license and began constructing his own aerobatic plane. He flew the plane and also helicopters and gyrocopters. Anything he could climb inside. He yearned to earn a living in the sky— somehow, some way—but his father had other ideas. He lured him into the family jewelry business with a promise of handing over the keys to the company someday. That day has yet to arrive; Juan's father is still not retired. Instead, Juan started his own jewelry-production plant and was soon thriving. He married Isabel at twenty in 1975, his dreams of flying deferred, but not forgotten.

"I have a dream to make a rocket belt," the balding Buddha says from the La-Z-Boy, his voice barely more than a whisper, here in the darkened workshop in the heart of Mexico. "And I have to see it through. My dream. So every free time I had, I was doing experimentation, and all the money I have, I invested in tools and machines to make the rockets." Once that meant having to pick up a small welding machine while the family was vacationing at Disney World. When the girls were young, Juan, Isabel, and the kids once piled into one of the minivans and drove to Orlando. They checked into the hotel after a twenty-hour trek. Isabel and the girls took off for Space Mountain. Alone in the room, Juan broke out the local Yellow Pages and found a machine shop that had what he needed.

Juan tells me all this in a steady, matter-of-fact way. The pursuit, the quest to fly, has become so entwined with his DNA that it is no longer amazing to him where his obsession might lead. This is a man, after all, who has spent nineteen years and God knows how much money just on the machine that can distill hydrogen peroxide to 90 percent so that he might use it as rocket fuel. Unlike most other would-be rocket men, Juan started with the fuel. It proved tricky. Very few companies were comfortable transporting the ready-to-use version of the stuff, and the ones that were willing charged an arm, a leg, and cash on top of that.

Juan decided it would be easier and cheaper to procure 50 percent peroxide from local chemical companies and distill it himself. (To put it in perspective, the stuff that pro ballplayers and male models use to turn highlights blond is only about 3 percent hydrogen peroxide.) This meant removing enough water from the liquid that the H_2O_2 was much more O than H. "You just want to boil off the water—and not explode!"

Sounds simple enough, but it turns out there is nothing simple about it, as Juan can attest. His first machine didn't get the job done. Neither could his second. Or third, fourth, fifth, or sixth. On the eighth effort, the grandfather with the gauze-covered legs nailed it. In nine hours, he can now distill enough fuel for a thirty-second flight. He's recouped some of his expenses by selling about twenty such machines to universities, hospitals, and fellow rocketeers for fifteen thousand a pop. "And there it is . . . ," Juan whispers, pointing to the glass-beaker, sci-fi-fridge thing I'd noticed before, lurking in the workshop shadows.

Aha—I glance at Jeremy to exchange a "Can you believe this crazy business?" look, but he is asleep, snoring quietly into his chest. "Perhaps we should pick it back up in the morning, Juan. It's getting late."

He nods and reaches for the phone. A quick exchange in Spanish— I kick myself for opting for high school French. "Your dinner is ready," Juan says. "In the house." I wake Jeremy, and we say goodnight, leaving Juan in the La-Z-Boy.

The food—yellow rice with peas, breaded chicken, and some unbelievably savory, lime-juice-slathered jicima—has a caffeinating effect on Jer. He talks animatedly with Isabel as I scope the house from the kitchen table. It is a modest two-story affair, but with considerable charm. Thick rivulets of ivy fall over one broad wall; a miniature waterfall cascades in a narrow pool, hugging the front of the living room, just beyond a grand piano, which Claudia uses to practice.

"When we were sixteen, Juan asked if I wanted to be his girlfriend," Isabel tells us. "And I said, 'Well, let me think about it.'" She laughs her big, toothy laugh. A cat races by the sliding glass door of the kitchen. "We have five cats—they are the enemy of my husband!" Isabel is a kind of one-woman SPCA, taking in strays whenever she can. A hand-stitched sign near the sink proclaims, "Where there is love, there is life."

After dinner, Isabel leads us around the backyard pool to our room, a small and cozy place with a slanting yellow ceiling and a lot of wicker furniture. She disappears and then quickly returns with clean towels, an ashtray, and bottled water. "Be sure to keep your door closed," Isabel tells us, "to keep the scorpions out."

As soon as Isabel leaves us, before I'm even sure she is safely inside the house, Jeremy turns to me and bellows, "Duuuuuuuuude!"

"Yeah?"

"Can you believe this?"

"I know—the food, the room, the pool, Isabel, isn't she great?"

"No, Juan. He's all broken, smashed to pieces. Been that way for the better part of a month. Just sitting in a chair all that time."

"Right," I say, not getting it.

"Well, you can pretty much sit in a chair anywhere you want—and where Juan Lockzumo or whateverthefuckhisnameis wants to sit is not in his bedroom with his wife, but up there. Alone. Guarding those machines. Dude has a bed up there—did you see that?"

Now that he mentions it, I did see that, right next to Juan's chair, within falling range of the La-Z-Boy. "Shit, that is crazy," I say, ignoring the fact that Jeremy wants to explore the phenomenon more deeply while lying down on my side of the queen-size bed. It's been a very long day. Did I really wake up to tickle face this morning? Was that only just this morning? I'm asleep before I realize it.

In the morning, I lean up on my pillow and look out the window. I see avocados, limes, lemons, lychee, and kumquats dangling lustily from drooping branches. A wall to one side of the pool is absolutely billowing with the most beautiful bougainvillea—purple and orange exploding fireworks of flowers.

After a pancake breakfast in the house, Jer and I climb the half-constructed concrete steps to find Juan exactly where we left him, in the same chair, wearing the same shirt, shorts, and slippers. In the day-light I now see the bike Juan wiped out on, parked in the middle of the shop floor, a mangled reminder. It could be any other two-wheeler, built for a ride through the Sierra or for fetching a bottle of milk. It's a white one speed. That one speed just happens to be 75 mph. And it has an 1,100-horsepower rocket and a torpedo of a fuel tank welded where the seat should be. And the back fender is crumpled like a soda can. Along the near wall, there's a bookshelf packed tightly with titles such as *Parachute Recovery Systems* and *Motorcycle Chassis*. On a cluttered desk, a stack of unopened mail competes for space with a motorcycle-hologram paperweight, foamy jeweler molds, and a photo of Juan's grandson.

Our host is flipping through a large stack of papers on his lap. He pulls one page out. "Did you know that rocket fuel is good for treating people with cancer and other diseases? And even the people with AIDS?"

I did not, in fact, know that. The page Juan has handed me is from a medical Web site I don't recognize; it's all about the medicinal benefits of peroxide. An elixir of life, it claims.

"I take every day eight drops of peroxide in a glass of water. So does my wife and our kids."

"How about your grandson?" I'm joking.

"Yes, we put it in his bottle."

Oh.

I don't want to think of Juan as some sort of Jim Jonesian cult leader, so I begin poking around the workshop, letting the conversation trail off. It is here, in this room that could barely fit two parked cars, where for the past ten years Juan has constructed his flying machines. He started with the catalyst, understanding that if he could make a more reactive pack, his fuel would be more efficient. At his jewelry shop he'd experimented with different metals. After two years of tinkering he'd come up with something he calls a "penta-metallic" cat pack. Or about one hundred skinny screens, their diameter wide as an orange, made from silver, gold, platinum, palladium, and rhodium.

Next came the corsets, which I see in the daylight are like sleek fiberglass haute-couture vests from a Milan fashion show, circa 2050. His wife and older daughter poured the plaster molds for the corset over his back. Juan treated them with epoxy, microballoons, and fiberglass until they were blast-shield smooth.

Then he turned to building the belt's mechanics with the arsenal of welders, lathes, and mills he'd amassed online, around Mexico, and during family vacations to Disney World. By that point he had acquired from a network of conspirators and like-minded pilots in waiting a good-size library of plans and diagrams and even the official Bell throttle-valve blueprint.

For a minute I wonder why, once he had all the tools, the parts, the years and years of planning, why after all that did it still take him eight more years to complete a belt? But then I pick up the stainless-steel shower-head-looking part that disperses the superheated hydrogen peroxide evenly into the catalyst chamber. There is a small Milky Way of pinpricked holes drilled into the face of it. I estimate there are about two hundred surgically made incisions.

"Did you drill every one of these holes?" I ask, trembling slightly at the mania.

"Yes."

So there's that. And the two years spent tinkering with the five-metal catalyst pack, and, well, everyone knows how sensitive the throttle valve can be. Plus, he thought to shorten the nozzles considerably from what the Bell team came up with, raising his center of gravity, making his machine easier to control.

I shuffle around Juan's workshop, shelf after shelf stuffed with bolts, wrenches, screws, tanks, and parts and parts and parts. Mexican soft jazz again tinkles from the cassette player. I run my hands over his plastic flight suits. A white one is torn and scribbled on with black marker, "Dad, I am very proud of you. I love you, Claudia," it reads in Spanish. Other relatives and friends have signed the suit.

"This is what I wore the first time."

About a year ago Juan suited up, strapped on a white crash helmet, invited a dozen of the people he loved most in the world over to the house, and marched them to the fenced-in backyard, where a twenty-foot-high safety crane shoots into the perpetually balmy air of Cuernavaca like an exotic metal pet. Juan hooked his white rocket belt to the tether. He kept his gray one on deck. His older daughter passed out noise-canceling headsets to everyone, including Grandma and her eighteen-month grandson. The boy, Jorge, manned a garden hose with Juan's wife. Just in case. The jeweler's lone assistant, a twenty-three-year-old rock musician Claudia knew from school, Rich, his dreadlocks tucked under a Rasta hat, raised a video camera to his eye.

Juan said a prayer under his breath and looked to his family, nerves conquering thrills. He cracked the throttle—it had been forty-two years since he first saw a man fly through the air at the Campo Marte campgrounds. The machine screeched with 110 decibels, waged war on gravity. And lost. The power just wasn't there. So Juan switched belts.

This time he soared—well, more like struggled slowly up, up, up. The hot steam squealing behind him. The grass blasted flat below. Seven feet into the air, hovering there for twenty-one seconds, before dropping down. A smooth landing. A beginning at once incredible and humbling. Later, a neighbor who'd heard the screeching vapor asked if he should be worried about explosions. "Everyone was very happy," Juan says now, shifting in the chair. "Me especially." He picks idly at an elbow scab.

Juan flew only twelve more times connected to the steel safety wire before he attempted a free flight. It was too soon—he landed too quickly and fell over with eighty pounds of rocket belt riding him into the loamy turf, breaking three ribs.

By then, the media had started paying attention. The BBC sent a crew, and so did a couple of local TV stations. Los Angeles morning radio producers were smitten. Introducing Juan on the *Kevin and Bean Show,* the hosts had this exchange:

Kevin (or maybe Bean): I am going to deliver to you now a guy who has built a jetpack. On his own. From scratch.
Bean (or maybe Kevin): Ooooohhhhhhh, no way! No way!
Kevin (or Bean): Juan, I don't know if they told you when they booked you, but I am obsessed with jetpacks and just am praying to live long enough to see the jetpack fly in my lifetime.

Whether it was the canned hysteria of a couple of morning zookeepers that had fired him up or the impatience that comes with thirty years of work, Juan was tired of waiting. Whatever the reason, a week after the *Kevin and Bean* interview, his ribs still mending from the rocket-belt fall, Juan flew again, back on the tether, for the Discovery Channel.

I ask Juan if I can try a belt on. He waves sleepily at the silver one— sure. He calls out to Rich, who bounds up from the tool room below us and helps me into the corset.

Despite the fact that at 230 pounds, Juan is considerably heftier than me, it's not easy contorting into the thing. I guess it makes sense that it's snug. After a brief struggle, I'm in. Rich lifts the tanks off a latch securing them to a steel stand, the easel displaying Juan's masterpiece. When he lets go, I feel it. The weight. It's no joke—and this is without fuel. Once this baby is gassed up, it'll be twice the load. It's all I can do to stagger in a drunkard's semicircle without my knees buckling. Still, I'm elated. I'm not even in a flight suit, just jeans and a button-down shirt I might wear any day of the week. But it doesn't matter. I'm a suburban dad in the year 2029. I wake up in the morning, take my coffee-bacon-and-eggs pill, implant the morning news chip in my brain, and blast off to work.

Suddenly, the cool fantasy crashes—Jeremy is on his back on the tile floor, clicking away. His fancy digital camera died at the airport, so he's forced to capture history with a cheap disposable one bought while procuring Juan a bottle of thank-you rum. "Good, and turn toward me."

I feel a little ridiculous. Juan looks only marginally pleased. I think he's worried I'm going to drop it. I look at my host with pleading eyes and tilt my head toward the backyard. "So. . . ."

"Yes?" Juan says.

"Mind if I take her for a quick spin?"

Juan looks at Rich as if to say, "Is he for real?" and then bursts out laughing. It is the most emotion I've seen him express yet. "No, that is not possible."

I'm deflated, but deep down I know he's right. Most rocket-belt pilots train for months, inching their way from a simple grounded throttle test to a two-second hover to tethered elevation. "Uh, Rich, can you help get this thing off me?" Though he doesn't appear to speak a word of English, the Mexican Rasta knows exactly what I'm saying.

Jer and I return to our plastic seats near the rocket-belt Buddha. Nino Amarena once sat here, seeking council on how to distill peroxide to make his ThunderPack soar. So had Troy Widgery and his Go Fast!

boys. In town to do a demo, they hadn't been able to get airborne. Not enough thrust. Juan shortened the nozzles, widened the fuel tanks, and—presto!—they had liftoff. Who knows how many others had occupied these chairs? Juan told us that years ago he had sold miniature rockets to an English company called Intura—he'd shown us clips of helicopters rising in front of frosty Swiss mountains—and to the Israeli army, also to up the power of their helicopters. This was ancient history as far as Juan is concerned. Just a deal between men, brokered over the Internet. He told us gentlemen from Iran and Iraq had gotten in touch, too. But he hadn't returned the e-mails. Now it's our turn here.

"I have a dream that is repeated constantly," Juan says. "I am flapping my hands." He demonstrates, making spastic flapping motions with both hands. "And it takes a lot of energy to get off the ground, but I am flapping and flapping and flapping until I go flying, and I enjoy it, going everywhere."

"Where do you go?" I ask.

"To see my friends and I go in the yard."

"Do you go to other countries?"

"No, just local flights." He laughs a deep rumble. "And the altitude, since I have flown gyrocopters, I have a very clear idea how the things look at low altitudes, so they are very believable dreams."

"Then you wake up."

"It's the worst part. When I wake up."

I wonder for a moment what Freud would think of Juan's dream. The thought is in my head before I can dismiss it. And here is what the maligned godfather of psychotherapy would say about such things:

The flying dream is the image which is found appropriate by the mind as an interpretation of the stimulus produced by the rising and sinking of the lobes of the lungs at times when cutaneous sensations in the thorax have ceased to be conscious: it is this latter circumstance that leads to the feeling which is attached to the idea of floating.

We leave Juan in the La-Z-Boy and walk through the garage to catch a taxi to the center of town in search of lunch. There, parked and nearing completion, is Juan's 22,000-horsepower motorcycle. The rocket he plans to install sits next to the bike. It looks comparable in size to the so-called Daisy Cutter bombs dropped on Afghanistan and seems plenty big enough to take out Rhode Island, were it designed for that purpose. Strapping the thing to two wheels and riding it brings to mind Slim Pickens's farewell journey astride a missile, yeehawing to his demise in *Dr. Strangelove.*

It's midday on a Saturday, and downtown Cuernavaca is bustling. Men with perfectly pomaded black pompadours amble the narrow, stone streets. Big-hipped women wrestle strollers over the curb. In the central square, an older fellow with the face of a well-made, well-worn glove sells giant inflatable junk. An oversized pencil balloon. A sickly sun. Dora the Explorer.

Jeremy and I get a little lost on the back roads. We pass butcher shops, newsstands, an Internet café, then a promising-looking restaurant—sturdy, dark wood tables, plenty of patio space under a tangerine canapé. Sealing the deal are a few plaques into the foyer detailing the building's history. Even with my less-than-zero Spanish, I am able to make out that the restaurant was once owned by the family of Anne Spencer Morrow, whose father was a New Jersey senator and the U.S. ambassador to Mexico, and who married history's most famous aviator this side of the Wright brothers.

Anne herself lived very much in the public eye, particularly after she met Charles Lindbergh. They were married in 1929, two years after he'd piloted the 233-horsepower *Spirit of St. Louis* on its famous thirty-six hundred–mile flight from Long Island to Paris armed with but five sandwiches "wedged between seat and fuselage" and a canteen of water. Did anyone understand the poetry, the promise of flight, as well as Lindbergh? In his Pulitzer Prize–winning memoir, he wrote, "Let a cylinder miss once, and I'll feel it clearly as though a human heart had skipped against my thumb. I push my fingertips against the quivering, drum-tight fabric of the cockpit wall. The plane's entire structure is carried by this frail covering of cloth. Thousands of pounds are lifted by these criss-crossed threads, yet singly they couldn't restrain the tugging of a bird."

Here, now, on the walls of this tucked-away Cuernavaca lunch spot are large framed photos of Charles with Anne and her family, a Con-gressional Medal of Honor pinned to a thick gray wool suit, his slen-der, dark-haired bride on his arm. The complicated horror of his life—the kidnapping and murder of his son—had yet to visit Lind-bergh's home when this photo was taken, and he looks strong and wise, capable even of building a jetpack. I'm lost in reverie.

"C'mon, let's eat." Jer pops the daydream. "I'm starving—and I could really use a beer, too. It's hot down here in merry old Me-he-co."

When we get back to Juan's place, his daughter Isabel is sitting with him in the workshop. Two months earlier, Isabel had become the first woman in the world to fly a rocket belt, pulling on the custom pink 'pack Dad had made for her and climbing smoothly a few feet above the ground. She hadn't realized she was going up that day—Juan had hastily arranged for a local television crew to come over in the morn-ing to capture history being made. "I flew at, like, noon," Isabel tells me, sitting near her dad's chair, coal-black hair pulled back into a pony-tail, revealing the same fine, sloping features as her father. "And then I was on the local news at three that afternoon!" The same toothy smile as her mom. "I had to fly and do the interview and everything quickly because I had to pick my son up at school at one."

I ask her if she was afraid.

"I was a little scared because I didn't know what I was going to do. It is—*oy*. How to explain it? First, it's very difficult to maneuver be-cause if you move a tiny bit, the controls, you go like a piñata. So as soon as you open the . . . *a qué?*"

Juan: "The valve."

"Uh-huh, you feel like something is pulling you up. You look to the floor and, ahhhhh, it's great, it's great. You are flying, and it's great to see your feet."

Isabel did amazingly well, especially considering she only had quickie instructions from her proud pops. Part of the reason Juan rushed her into the sky is he'd heard that Ky Michaelson was making plans to have his wife, Jodi, become the first female rocket-belt pilot. Among the generally goodwilled Yahoo! group members, one notable exception remains the often prickly, highly combative relationship between Juan and Ky. ("I don't like him because he's a big mouth. He brags about work he has never done," Juan confided to me yesterday, but eventually they would work things out.)

Though she had been rushed into her first flight, Isabel had had a lifetime of her father's dangerous games to prepare her for the big day. As a toddler, she rode his shoulders while he skidded across a nearby lake on a Jet Ski. By the time she was three, she was out on the water by herself. Dad built his girls the fastest go-carts in the neighborhood and taught them to fire pistols and rifles. Later, Isabel and Juan would take two of Dad's four BMW motorcycles out for rides together. Spring and summer meant hang gliding; winter was for speed skating at the local ice rink. Amazingly, Isabel grew up mostly unhurt, but on one motorcycle outing an errant car door forced her and Juan to skid across a patch of hilly pavement. She only split her pants, while Dad broke a finger. It wasn't nearly enough damage to dissuade her from letting her son later ride with Grandpa, giddily straddling the gas tank, a mini helmet rattling on his little head.

Besides the BMWs, over the years Juan has owned and operated, often with his daughters in tow, an aluminum aerobatic plane he built from scratch, a Formula 1–style racing car, four hang gliders, a gyro-copter, two racing go-carts, a jet boat (named *Cafita* after his father's nonsensical, all-purpose exclamation), a four-wheeler, plenty of cars, bikes, and ice skates of all kinds—inline, hockey, speed.

Juan and his daughters come by their extreme-adventuring appetite honestly. His grandmother, he tells me, was the first woman in Mexico to carry a gun and wear pants (both strike me as difficult to fact-check). Carmen Gallo was also, apparently, the first Mexican female to ride as a passenger in an airplane. She still drove a car around Mexico City when she was ninety-six years old. When she was younger, she won shooting contests and rode horses.

Over time, Juan sold many of his toys and poured the money into building a rocket belt. The hang gliders were difficult to part with, but after one serious scare, he knew they had to go. Eight years ago an old flying friend and he were climbing, soaring, riding the air, when suddenly a cumulonimbus cloud loomed overhead. Juan knew how to deal with the threat, turning his glider sideways and piercing the octagonal puff. His friend was not so lucky. The cloud current sucked him straight up, ripping him to twenty-five thousand feet, where he died of hypothermia and a lack of oxygen, his lungs bursting under pressure. Juan sold the family's gliders and hasn't flown one since.

This seems like the right time to ask Juan about death. Well, as right as any. I ask him if he ever thinks about how he might go.

"In an accident. I crash and then that's it—fast, easy, and without pain."

Isabel scrunches up her face—"Oooohhhhh, no, no, no, no, no."

"I don't want to die of cancer or something like this," Juan continues. "That's why I told my daughters and my wife to let me do the things I like. I was reading about a man named Jerry McBride—he rode motorcycles all his life. He was a champion; he rode over 200 mph, every weekend. And he died of a kidney implant, because the dialysis was not successful, and he died. And Cy Miller—he is the fastest man in the quarter mile racing a rocket car. He rode over 400 mph, and he was killed in Texas when the hook of a crane broke off and hit him and he was decapitated. He raced for more than thirty-five years. It's absurd, but it happens."

The room is quiet; Isabel looks ashen. "He thinks that because my sister and I are old enough and I'm married and make my own family and Mom is okay and everything that we don't need him anymore."

"No," Juan counters, "that now I can do what I want. I want all my life for them to have a house, to have the best education, have everything. Now they have their own lives, so I can make what I want."

"I'm not prepared to lose my father."

"You will someday."

"Yeah, but he's so young," she looks at me, pleading, it seems, hoping I can talk some sense into her old man. "He's only fifty-two."

"I am becoming older very fast." The rocket maestro groans slightly, or laughs, and shifts painfully in his La-Z-Boy for emphasis.

Before Juan gets too old, he hopes to complete his fleet of six rocket belts, to train a few pilots to fly, and to begin making a killing at local football matches, with sponsorship by Pepsi, Coke, Budweiser. Time would seem to be on Juan's side. The youngest any of his relatives died was ninety-three years old.

"That's why we are so angry." Isabel won't let it go. "Because we don't want him to prove anything else. And he has never had an accident like this." She waves a hand at the gauzy bandages, holding beneath them the flesh he had removed from his calf and grafted to thighs. "Never."

It's getting to be late in the afternoon. Jer and I are due in Mexico City this evening to rendezvous with three horror film fans my traveling companion met during a festival last year. In the morning, we'll fly back to Los Angeles.

Good-byes are drawing near. Before they arrive, I say, "The lesson you seem to want to teach your kids and your grandson is: Don't be afraid."

"Live your life and do what you want, and at the end the only one that knows the way you will die is God." Juan is staring past me, out the open door of his workshop, out past the avocado trees, the lemon and lime trees, the lychee plants and beautiful bougainvillea, past the pool to his fenced-in backyard, where the tether awaits his next attempt to fly. "Maybe after doing all I have done with my life, I will die in a hospital bed. I don't know. That would be a shame."

I don't want to leave, but there is nothing more to say. I'd like to stay—a spasm of a daydream flashes in front of me. I call Catherine, tell her to pack up the kids and some clothes and come down here. Forget the L.A. wedding or the flight back east to face another brutal winter indoors with kids wishing they could break free and roam. Come south, take off your shoes, dip in the pool, and drink tequila at the built-in aquatic table, plucking avocados off the branches in the land of eternal springtime.

But then we really are saying, "Good luck" and "Thank you so much for everything" and "Feel better," and making plans to come back soon. The Discovery Channel may want to shoot Juan again; perhaps we'll return then. I find it difficult saying good-bye to Juan's wife. She's been so good to us, after all, showing us around and feeding us while her husband's been stuck indoors.

Jeremy and I step blinking into the sun. Down the half-constructed concrete steps—an awkward, wordless wave to Rich, who is sanding the fiberglass husk of the rocket cycle as if his very life depends on it. In front of one of the student rooms, a bearded middle-aged man in cargo shorts and gingham shirtsleeves is sitting at a picnic table, consulting a Spanish textbook. A woman who could be his mother relaxes in a chair next to him under her tight perm. They are from Dallas, they tell us, back again to study Spanish after enjoying themselves so much last year. And the year before. They've grown close with the Lozano family. Mom does the shopping with Juan's wife. They were here when Juan flew for the Discovery Channel with three broken ribs. I ask what they make of their eccentric host.

"My Lord," the son says, his vowels twanging. "I wasn't sure what was going on. But I do know that that motorcycle over there"—he points to the massive rocket in the open car park—"if Jay Leno ever sees one, he's gonna want it." The son pulls back from the joke to try to make sense of everything. "Mexico was basically a third world country, but now it's gone from the nineteenth century to the twenty-first century just like that." He snaps his fingers. "And if Juan Manuel is any indication, they could be a real powerhouse."

It's an interesting interpretation of the slice of paradise I've seen this weekend. I'm not sure I entirely understand what the Texan even means, but then I settle for an explanation I can live with, which is that Juan Manuel will be making us all jetpacks before the end of the twenty-first century. So all I have to do is hold on another ninety-four years!

And then we are back on the bus, back over the Sierra, dipping into the bumper-to-bumper smog of Mexico City. We meet Jer's three fans—Manolo, Lalo, and Felipe. They make a great trio. Manolo is dressed all in black and is roundly cute in the way that the Cure's Robert Smith is cute, works as a dentist, and keeps a rotating cast of black tarantulas in the back of his office. Lalo, his lover, is also dressed head-to-toe in black, a leather jacket over a T-shirt depicting a man with a halo or maybe something more ominous hovering over his head. He is muscular with thinning hair and works as a pharmacist, is writing a novel, and has a pet python in his apartment in a suburb north of the city. Felipe is tall, really tall, and bearded with long, stringy hair and an excellent crooked beak of a nose. He is a politically active actor who recently appeared in the film *Midnight Virgins,* and also participated in a protest of the recent presidential elections by helping close down the city's main thoroughfare for forty-nine days.

We drop our bags in the small, smelly, bloodred room we've booked in Zona Rosa and walk to a bar that Felipe tells us is popular with "intellectuals." Intellectuals in Mexico, it seems, take to dark, labyrinthine bars where pesos must be exchanged for fake Bar Milan bills in order to pay for drinks. And they also dig '80s New Wave bands like Siouxsie and the Banshees, and the Smiths, which dominate the jukebox. That's fine by me. I'm suddenly back in high school, pounding Indigo beer and shots of sagrita (like tiny Bloody Marys), nodding to the three beaming foreigners as they shout over Souxie's "Cities in Dust" about the brilliance of the Chilean director Alejandro Jodorowsky's horror oeuvre.

I'm exhausted, but Jeremy is just revving up. The languid afternoon at Juan's place was very much not his speed—he's used to something a little faster. Several drinks later, he wants food and to go "looking for something odd." I don't think he knows what he wants, but has an innate sense that this city is capable of producing strange sights—vulgar, leering, sinful priests; killer bugs; midnight virgins! Anything you might want. My friend truly sees the world through a horror film scrim and believes black magic can be found if you know where to look.

So after a cramped cab ride and hungrily scarfed tacos, we go looking. This leads us to a cobblestone square miles from the bar and hotel. Within seconds of tumbling out of the taxi, a midget in a filthy sweater vest and pants so long the hems drag like frayed dead rodents on the pavement squires us to an open-air bar—or, rather, a card table set up at the edges of the square suddenly with a full array of booze. On either side of the table, several similar entrepreneurial stands are doing brisk business with the downtrodden and desperate. There is a fee involved for the small man's work, I think, though that is certainly lost in translation. In any case, we begin stupidly downing shots of rum. In the center of the square, a roving four-piece mariachi band plays on and on—tum-da-da-dum, tum-da-da-dum—while a small pack of drunken, fleshy couples dance closely, heads pressed to chests, eyes closed, hands, nearly all of them, clutching sweaty cans of Modelo. I see intimate packs of dancers and bands scattered in the shadows nearby.

Now another small fellow approaches—whether he is a dwarf or just short is hard to say. He holds what looks like an old-timey cigarette-girl box, but Lalo reaches in and reveals that the innards are made up of two copper joysticks attached to an electric volt. He hands Manolo his leather and grabs a battalion in each hand, a smoldering Marlboro filter wedged in his teeth. "OK, go!" he shouts, and the little man twists a hacked, sinister dial. Unknown quantities of current shoot into the pharmacist's system. He holds tight, forearms twitchy, gnashing the fiberglass of the filter. The dial is cranked. Lalo hops on both feet but squeezes the rods still. Mercifully, the little guy shuts it off.

"Oh, you gotta try that!" Lalo's eyes are blazing, his face a roaring pinball session with an extra ball. Beer, sagrita, electricity—a wicked cocktail.

We all form a circle, holding hands, the most evil prayer group of all time. I have no idea why I am doing this, except because I am in Mexico City with Jeremy and the hell with it, anyway. The dial is done. I writhe with the jolt, a knife stabbed under the veins of my wrists. I last about as long as my American friend—three seconds, maybe, tops, and we throw our hands free of the pain. "What the fuck!?"

Lalo shrieks with delight, while his lover affectionately rolls his eyes. Felipe is bemused, drunk.

We pass up an overpriced strip club on the square despite heavy solicitation from Sweater Vest and make the long walk back to the Zona Rosa hotel. It's a serious hike, maybe five miles, along the shadowy, broad boulevards. Every now and then, a majestic marble building rises out of the pavement. I'm too drunk and tired to bother asking about them.

By the time we reach our home for the night, in fact, my bones are aching with weariness. But Jeremy, who is practically single these days, is ready to press on. Lalo, Manolo, and Felipe beg off—they have the good excuse of work the next day—so I have little choice but to accompany my old friend on a 3 AM hunt for a strip club.

Soon we are in a cab, the city swirling by in running acrylic colors, and then in a dingy, dark, busy club, the taxi driver waiting—or so we

really, really hope—outside. The staff immediately preys on two obviously sauced gringos, and before we realize it we are seated at the
foot of a pathetically short stage paying a few hundred pesos for
drinks we never ordered and whispering senoritas who either don't
understand us or simply won't get off our laps. In the severe black
lighting all this surrounding lavish flesh glows grotesquely phospho-
rescent. Now we are drinking vodka—mixed with the lights the world
becomes a Jackson Pollock of freckles and green vampire teeth. I don't
feel very good. "Oh, papi, don't you want a dance, papi?" Not really—
I just want to go home.

In the car to the airport the next morning I'm worried about Jer. Normally capable of rapid-fire, often hilarious narration at the worst moments, he is groaning, his head lolling on the taxi's crumbling-foam
seat back. When we finally swerve into the drop-off lane, he bolts from
the backseat, steps onto the curb, and lets loose an impressive fire
hose of liquefied taco from his mouth and nose. Then he straightens
himself up, wipes the back of his hand across his lips, and announces,
"Much better." Not one of the passing citizens of Mexico City seems
to notice.

Back in Los Angeles, I reunite with the girls at Nancy and Rob's
prewedding party in a vast manicured backyard in Santa Monica.
Though I've been away from them for only two days, it seemed much
longer. I can't quite bring myself to let go of Oona's hand all night,
which is fine with her since she needs help feeding the inflatable alli-
gator floats in the pool.

The day after the wedding I drive the rented minivan up through
Hollywood, onto the 101, and into the valley. To get there I pass very
near the neighborhood I once lived in, after ditching NYU for a go at
the movies. The streets are familiar: Detroit, Wilshire, La Brea. Soon
I'm pulling onto the stretch of Van Nuys Boulevard that houses Nelson Tyler's business. Wedged between legit auto-body-repair garages

and what might be chop shops, Tyler's sprawling warren of airy rooms for machining, electronics, and administration remains at the same address it was back in the early '70s when he first began constructing his own flying machine. The massive steel beams that held his pulley-operated safety tether still stand on opposite sides of the parking lot.

He greets me today in the lobby in a short-sleeved polo shirt, jeans, and unblemished New Balance running shoes. Where he once had a wild Einsteinian tangle of hair, it has retreated into a white, fluffy horseshoe. Tyler has a sinewy runner's physique with the body fat of a wrench. It's stretched across his six-foot-two frame. He possesses an angular, pale face and the energy and enthusiasm of a hyperactive six year old. He leads me on a tour of his facility, bounding ahead of me on his toes, staffers whizzing happily past us, while he leaps from topic to topic, with barely a moment for me to catch my breath and make sense of what he's just said.

Over the next couple of hours, I repeatedly try to return our conversation to all things rocket belts and jetpacks, but he'd rather talk about the present. Granted, I can see why he's excited about what he's up to. When it comes to camera mounts now, he'd just as well let the forty or so that are out there in the world capture helicopter pans and boating tracking shots without him. He'll just collect the rental fees, thanks very much.

These days, Tyler Camera Mounts is in the law-enforcement-supply game. Specifically, inside this well-lit Bat Cave in the valley, his staff of twenty turns out a myriad range of aluminum platforms that hook onto a helicopter's side, providing gunmen with a steady perch. His clients include the CIA, the FBI, SWAT units, border-patrol cops, assorted local sheriffs all over the country, and even some foreign governments. In the age of terror, business is good. "The FBI says they can't shoot a sniper rifle from a helicopter, unlike in the movies—so I built these."

The other thing on Tyler's mind is his costume—today is Halloween, and he's going as the Rocketeer to a party at his friend Nick Rush's house in Beverly Hills. Tyler's wife of sixteen years, Liz, will sport a cartoony 1940s outfit as the love interest, Jenny Blake. Rush is a writer-producer-director, probably best known for the 1980 Peter O'Toole film *The Stunt Man.* "He's my best buddy," Tyler tells me (thus his cameo in the movie).

In order to get his Rocketeer gear just right, Tyler's had his receptionists trolling the Internet for weeks. The result is a flawlessly detailed re-creation of the getup that remains a cultural touchstone for virtually every would-be rocket-belt pilot and jetpacker. The jacket, a caramel leather number that's part World War II foot soldier, part elevator operator, came from a Canadian company that markets them as the Rocketeer Jacket. The gun, resembling a *Star Wars* blaster, is due in the mail any minute from Florida. The boots were tracked down at a nearby police precinct. The helmet, which memorably made Bill Campbell look like a hood ornament, came in a kit for nine hundred dollars.

"Do you know what rotomold is?" Tyler asks, pointing to the finished helmet, lying on one of his work tables.

"You build a pattern with something, and you make an aluminum casting, and you split it and open it, and you put in some plastic powder, and you close the mold, and you put it on an arm which is rotating it, and it goes through an oven, and it's rotating, and the plastic melts and gets on the sides, and if you want you can put a lot of plastic— they use a little—and it comes out, and it continues rotating, and it cools in a cooling area, and then you pop the part out. And that's what that was, but of course it's closed, and it will have a seam, and they cut this, and you can see what it looks like—this is what's the melted part, and this is the same thing—so it just comes like a big, like, glob, and we did all this in a week and a half."

And then he bounds off, into another room, another story.

It is, in part, because of this dexterity of thought that Tyler has been able over the years to build just about whatever he can dream up. A superpowered Jet Ski? Sure, no problem (Roger Moore rode Tyler's model as Bond in the *Spy Who Loved Me*). A blimp bigger than Goodyear's? Why not? Now he's pointing to his latest creations, a remote-controlled bowling ball and a kid's scooter that he says will put Razor's to shame with its dual-steering-enhanced smoothness of handling.

"I am licensing the bowling ball out to Columbia, the biggest ball manufacturer. Bowling, by the way, is the biggest participating sport in America. And they've done four things in their history that are new. In the '50s they made the automatic pinsetter, you know, so you didn't have to wait for the guy to set them up. About ten to fifteen years later they did automatic scoring, which is always a problem for people. The next thing they did was the bumpers that come up for kids. Then they did cosmic or extreme bowling, where they turn the lights down on the weekends and the music up, and everything glows. Anyway, I got the idea for an electronic ball—when you roll it, you can steer it. Or I can roll it and you can steer it. You know, teams. And when you steer it, it flashes; it has lights flash inside, and it has running lights so in the dark you can see it and know where it is."

"Probably a good innovation for the video-game generation," I say, my head reeling a little.

"Yeah, that's what I thought."

Tyler's stream-of-consciousness rambling can lead to an odd effect. As we race up a flight of stairs, me lagging behind again, he says, "My electronic guy last year came in about six o'clock and said, 'I don't feel good.' And he died of a stroke at forty-two years old with two kids, and so we have a new electronic guy, but he comes in at nighttime on the weekend." This information has been imparted to me in such a nonchalant way, as if Tyler has just asked what I'd like for lunch, that it takes me a couple of seconds to realize he's just let me know that an employee of his is now dead. Perhaps he struggles with feelings that can't be leveraged into *things*. The tour goes on.

There's a poster of his wife, Liz, in the Terminator-like steady-cam suit Tyler designed. Sleek, black Kevlar vest attached to an articulated arm, at the end of which the camera is held. Liz looks like a film-crew army of one from the future. Though she doesn't don the suit much anymore, she once worked constantly—Tyler seems most proud of her role on the set of Stanley Kubrick's last film, *Eyes Wide Shut*.

Another high-ceilinged warehouse room. "This is my rocket-belt room," Tyler announces, speed walking through the door. There are posters from *King of the Rocket Men* and another *Flash Gordon*, which my host has ripped from the pages of pulp magazines, blown up, and framed. A cardboard-and-foil model of a flying rocket belt and pilot that Bill Suitor made for Tyler dangles from the ceiling. The inscription reads: "Nelson, one brief moment of glory is worth an age without a name. Thanks for the moment."

In the main workroom, the 747-long walls are crammed with posters from Tyler's past, movies like *True Lies, Executive Decision, Broken Arrow, Patriot Games, Lethal Weapon, Top Gun, Die Hard, Waterworld, Apocalypse Now, RoboCop 3, Firebirds, Jaws,* and *Rambo.* If an airborne tracking shot has been committed to film in the past forty-odd years, chances are Nelson Tyler has played some part in it.

We retreat to his office, passing three *Rocketeer* posters and two books about the movie on the way. Here the walls are devoted to *Metropolis, Blade Runner, Flash Gordon,* and *The Lost Boys.* On his desk is another photo of steady-cam-equipped Liz and two tiny Rocketeer models. On the coffee table in front of us is a stack of photo albums and the 1961 aeronautical engineering book Tyler consulted to build his belt—it's in remarkably good shape; the only damage seems to be the many dog-eared pages, like the one on how to mold nozzles and another about hammering out a catalyst pack. We flip through a couple of scrapbooks and see Tyler's two kids modeling the Jet Ski in ads;

Malcolm Forbes on a blimp ride; Soichiro Honda, giddy with the rocket belt on his back; and, of course, Bill Suitor at the 1984 Olympics.

Tyler was standing with Suitor on the top coliseum step when the pilot made the most famous rocket belt flight in history. "I tuned it, adjusted the pressures and everything. And then got my camera out."

While we're on the subject, I ask the restless genius what it will take to finally get our long-anticipated jetpacks. He didn't sound very optimistic at the convention last fall, but I wonder if here, on his own turf, he'll give me the real scoop.

"Oh, well, the main problem with the rocket pack is the twenty seconds, and the jet would probably be six or seven minutes, with any fuel you could walk around with. I mean, Lockheed [Martin] could do a good jetpack, and they would probably spend two hundred million, but Lockheed could do it. I could do it if somebody gave me six or eight million."

"Is that right?"

"It just takes lots of money," Tyler says.

"So, have you ever seriously pursued that six or eight million?"

"No, I haven't chased it. We've been so busy. This is our water tank, it goes all the way back there, and it is four feet deep, and of course it is obsolete, we don't need it anymore."

"You never tried to build one—a jetpack?"

"No." It's becoming clear he wants me to forget it. "There is nothing available, as far as the jet engines; all the little model engines are way too small."

"So why do all these guys. . . ."

There's a knock at the door—a FedEx package has arrived. Tyler tears into it.

"My gun!" He leaps off the leather love seat, racing out of the room. "My gun came! My gun came!" Tyler is practically singing, showing off his Halloween costume's final accessory to any staffer who'll stop to listen.

It's time to go—in a few hours I'm taking Oona trick-or-treating near a friend's place in Laurel Canyon. She's going as a genie—maybe she can make a jetpack appear. Tyler's staff is skittering about, prepping for a catered holiday lunch.

We walk out to the parking lot, and we both instinctively crane our necks to take in the soaring A-frame that held the tether Tyler used when he decided to make history.

"Okay, well, be in touch," he says in his good-natured chirp. He's a curious creator, a lovably unbridled, eternal tinkerer.

"Will do."

I turn and walk out onto the street to where the minivan is parked. It's only as I unlock the door that it occurs to me that I'm standing about half a mile from the Van Nuys courthouse where Larry Stanley and Chris Wentzel went on trial for the kidnapping and torture of Brad Barker. It happened almost exactly four years ago, November 22, 2002, marking the freaky denouement to the southern Gothic tale that was the blood-soaked disaster known as the American Rocketbelt Corporation.

CHAPTER
6

Houston, We Have Another Problem

He who binds to himself a joy
Does the winged life destroy
But he who kisses the joy as it flies
Lives in eternity's sun rise.

**—William Blake,
"Songs of Innocence and Experience"**

Bradley Wayne Barker looked like a Hollywood cop. Or maybe a utility infielder for the Mets, circa 1987. He cultivated a dark, feathered mane of hair, had a jaw broad as a garden hoe, and willing dimples. Barker also had the Clooney-like charisma to match. "With his smile and good looks, he made a charming first impression" is how *GQ* magazine described him in a 2002 article. "Brad can be very charming and comes across quite well and as though he's educated," Nancy Wright, an old acquaintance, told me.

If she sounds dubious, that's for a very good reason. There was a time when Brad Barker was the most likely suspect in the brutal 1998 murder of Nancy Wright's brother, Joe. At the time Joe and Brad were partners, along with computer and aeronautics entrepreneur Larry Stanley, in a Houston-based outfit called the American Rocketbelt Corporation. The idea was to build a belt like Nelson Tyler's and earn about twenty thousand dollars per thirty-second demonstration— not a bad day at the office. But things didn't really work out that way.

Joe's murder, though certainly the most tragic event stemming from the ill-fated endeavor, was only one of several bewilderingly strange and often violent moments in the brief but raging history of ARC. Although the case has never been solved, its place in the dark stratospheres of Brad Barker and Larry Stanley illuminates what happens when an obsession with flight is mixed with greed, jealousy, vengeance, money, and enough powder-keg personalities. Some jetpack dreams, it turns out, can swerve into real-life nightmares.

Born on June 14, 1954, Barker had a fairly typical middle-class childhood, growing up in the heart of Illinois. But when he was eight years old, his universe imploded the moment his dad died in a car crash. Soon after the accident, Barker, wandering the streets in a fog, found himself in a movie theater. The featured attraction? *Thunderball*, starring Sean Connery and his jetpack getaway. "I was in a daze," Barker said years later. "But that rocketbelt mesmerized me."

It would continue to do so through many years, many moves, and several dead-end jobs. An itinerant searcher, Barker drifted aimlessly through his teenage years, finally putting down roots in Houston in 1975. It happened quite by chance. He wandered into the Central National Bank looking for a job—any job—and when the man who interviewed him turned out to be a kindred spirit, he was hired to sell insurance.

His new boss was a similarly restless soul named Kinnie Gibson. The two young men became fast friends. Among their dovetailing hobbies was a deep love of airborne thrills, which they sought by flying Barker's Cessna 210 Centurion and by skydiving. At night, flush with bachelor cash, they cruised the downtown singles bars, outlaw swaggers deployed to seduce southern girls glistening in Houston's swelter. When they grew up and got married, they remained just as tight. Barker was the best man at Gibson's wedding, and Kinnie returned the favor. As English journalist Paul Brown notes in his book about the ARC fiasco, *The Rocketbelt Caper,* the two men's sons were born within twenty-four hours of one another.

So when Gibson lit out for Hollywood in the early '80s and eventually found his fortune (a reported $1.4 million), blasting off with a rocket belt during forty performances of Michael Jackson's European tour in support of his album *Bad,* Barker was decidedly bummed on two accounts: his closest friend was no longer around, and while Gibson was glamorously cavorting with celebrities, Barker slogged through a series of odd jobs while his marriage unraveled.

Then, in 1990, Gibson invited Barker to join him in California, unwittingly setting in motion a series of bizarre events. By that point, Gibson had done his time flying Nelson Tyler's rocket belt. After Tyler unloaded his flying machine to Copenhagen's Tivoli Gardens amusement park, Gibson swooped in to buy the belt himself. Now he needed a crew; that's where Barker came in. He followed his old friend first to Los Angeles, and, when Gibson was hired by Disney World to fly daily exhibitions, Barker followed Gibson to a place well steeped in unusual happenings: Orlando, Florida. For a while the two old friends seemed to have worked out a professional arrangement: Gibson launched himself heavenward and was paid handsomely to do so; the money trickled down to Barker.

But loyalty goes only so far in certain business deals, and soon enough Barker began chafing in his subordinate role. One day a visitor showed up in Orlando, a new friend of Barker's named Joe Wright who owned a thriving car-stereo company in Houston and had once hooked Barker up with a sweet system. Wright was a slim and fastidious guy (he kept a notebook for jotting down jokes he liked), partial to ironed jeans and Doc Martens. He was also a closeted gay man with a taste for crystal meth.

If Gibson was uncertain about what exactly Wright was doing in Orlando, that was cleared up soon after his arrival when the rocket man stumbled upon Barker and Wright videotaping the belt. Seems Barker had flying ambitions of his own. When Gibson's throttle valve spit out a busted piston, Barker saw his chance and seized it. Gibson dispatched Barker to Houston to repair the valve with machinists familiar with the part. Barker did as he was told, but not, according to sources familiar with the events, before measuring the parts and detailing the specs. Years later Larry Stanley noted in an online statement about those events that "[Barker] believed he had acquired the great secret behind the belt's operation."

That may or may not have been true, but not long after his Houston sojourn, Barker formed a 50-50 partnership with a friend he'd met through Gibson during a skydiving outing. In 1992 Brad Barker and Larry Stanley founded the American Rocketbelt Corporation. Stanley, a beefy man with a *Magnum P.I.* mustache, lived in the affluent Houston suburb of Sugar Land, founded as a plantation in 1838.

Barker estimated the costs to get their own machine up and flying to be about two hundred thousand dollars. Stanley, whose family owned a small oil field in Missouri City, would later claim that he pumped about half that into the nascent business, while in court testimony Barker estimated that he sank about eighty thousand dollars of his mother's money into the operation. And there was a third, mostly silent, partner. Stanley has said he believes Barker squeezed roughly fifty thousand bucks out of Joe Wright, while promising 5 percent of the profits from the flying belt's exhibitions.

Barker also convinced Wright to let ARC set up shop in a corner of Wright's car-stereo headquarters, wedged into a busy stretch of commercial real estate next to the Trophy Club strip joint. And there, in 1993, work began on what would become the Rocket Belt–2000, or RB-

2000. The hope was that it could utilize modern materials like tita-
nium, aluminum, and stainless steel to be a lighter, and thus longer-
flying, version of Nelson Tyler's rocket belt—and that it would bring in
several hundreds of thousands of dollars a year.

Stanley and Barker were unlikely partners, to say the least. Only
three years earlier Gibson, while in the Philippines doing stunt work,
had asked Barker to drive out to the Stanley family oil field to recover
some tools and machinery of his that he believed Stanley had stolen
from a storage facility. Stanley and Gibson had of late been bickering
over an oil deal gone sour. Their friendship was on the rocks. Barker,
rarely prone to subtlety, grabbed a baseball bat, called up a karate-
instructor buddy named Rob, and sped out to Missouri City. There,
they were met by a Stanley employee, a former Navy SEAL, according
to Paul Brown's book. "Rob went into his karate routine," Barker told
Brown, "and in literally three or four seconds, just beat the shit out of
this Navy SEAL." That didn't stop Barker from piling it on, and he allegedly cracked the hapless SEAL on the legs with the bat, while repeatedly shouting, "Where's Larry Stanley?" Yes, Barker could be
charming, but he could just as easily be a violent menace.

This is about the time Stanley pulled up, saw what was happening,
and decided to return Gibson's gear. Strangely, as Brown noted, while
Stanley loaded Gibson's tools into the car, Barker said nothing about
his Cessna. Four years earlier the beloved plane had gone missing after
Stanley borrowed it, only to turn up in a Seattle hangar, tricked out
with thirty thousand dollars worth of upgrades. Barker couldn't afford
to pay for the new plane parts and was forced to sell the Centaurion.
He was decidedly pissed and blamed Stanley for costing him his baby.
Before that day on the oil field, the two men had not spoken a single
word in almost half a decade.

And now they were partners in the fantastically alluring, if virtually
impossible, flying game. So perhaps what happened next isn't a complete shock. As the RB-2000 prototype moved toward completion,

Stanley began to suspect that Barker was embezzling funds by claiming machine work cost about twice what it really did. His fears were confirmed when one shop owner contracted by ARC told Stanley that work Barker said cost eleven thousand was actually only fifty-five hundred dollars worth of parts and labor. "In other words, I was buying all the parts," Stanley later told the *Houston Press.* "Barker wasn't paying for anything."

The jilted partner was swiveling in an office chair at Car Audio Plus one day, detailing his grievances against Barker in a phone call to Wright, when things came to a head, so to speak. Stanley was unaware of Barker lurking in the shadows, eavesdropping on the conversation, his notoriously hot temper pushing 1,300 degrees Fahrenheit. Fuming, and maybe also somewhat ashamed, Barker picked up the nearest blunt object—a four-pound, hunter-orange, lead-filled mallet used for pounding out dents on auto bodies.

Earlier that day, in a money-related brawl, Barker had broken a finger on his left hand when he and Stanley had crashed through an office door. Now, a sling on one arm and the mallet in his good hand, he lunged at Stanley, repeatedly hitting him on the skull. Stanley put his right hand up to deflect the blows and saw the lead land with a dull thud on his ring finger—tendons snapped, the digit went limp. Stanley would later tell the *Houston Press* that Barker was screaming, "I'm going to kill you, motherfucker!" When a machinist grabbed Barker, hoping to break up the fight, he was greeted with a blow to the knee. Finally, a Car Audio Plus employee was able to temporarily quell the violence. But before the dust had settled in the warehouse, Stanley and Barker were back at it in a furious fistfight that ended with Stanley lying in a heap, his white oxford shirt stained cranberry from a gash on the back of his head. "I stood over him and told him to stay on the ground," Barker told Paul Brown. "And he did."

The fight was over, but the trouble was just beginning. After a trip to LBJ General for stitches, a CAT scan, and to have his finger repaired,

Stanley was dumped in the Harris County Jail. His heart sank when he saw his cell mate, Bradley Wayne Barker. The pair made for quite a sight, Stanley bandaged about his head, Barker's arm in a sling. They were each charged with assault and released that afternoon.

Stanley spent the following week in bed, recovering. Barker, meanwhile, was much busier. He and Wright hired a lawyer who issued a twelve thousand–dollar lien against Stanley for unpaid rent in the stereo shop. In a public auction, the new partners bought all of ARC's assets, including the RB-2000, for ten grand. Barker drove to Wright's shop and happily loaded his car with the belt, a homemade hydrogen peroxide fuel distiller, and, according to Brown's book, a .357 Winchester rifle that belonged to Stanley. With partners like these, not having a gun was starting to seem like bad business sense.

By now work on the belt had progressed to the point that a first test flight was on the horizon. Barker had ironed out early mechanical problems with the help of California aerospace engineer and inventor of the forty-foot-high fire-breathing Robosaurus, Doug Malewicki. "Barker called me from Texas one day—he had questions about the catalyst pack," Malewicki told me in a phone conversation. "They paid me for a couple of weeks' work that a college student could've done."

As a testament to Barker's ability to charm, Malewicki's first impressions of the man on the other end of the line were that "he seemed like a nice guy, a friendly guy." But Malewicki admitted that "maybe I'm not astute at picking up people's ulterior motives." Then he noted that, from what he could tell, "Barker wanted to be the pilot; he wanted to be the star." There was only one problem with this last desire: the would-be airman had no idea how to fly.

Enter, once more, Bill Suitor. Perhaps it was the money. Or the fame. Or perhaps Suitor just really loves taking to the air. It's hard to say exactly why a man who had done it all where the rocket belt was concerned would agree to get mixed up with the likes of Brad Barker (and he declined to discuss the issue with me), but it was a decision Suitor would come to regret.

When Barker called, Suitor was living a quiet life in Youngstown, not far from where he was born. He was nearing retirement from his job as plant operator at the power authority. In his free time Suitor carved and painted duck decoys in a studio behind his house—it was a hobby he cultivated as a boy and still adored. But when it came to flying, he couldn't say no.

Malewicki says he helped design the RB-2000's engine for a 175-pound pilot, but Suitor arrived in Houston tipping the scale at close to 230. A welder who worked on the belt told *GQ* that this irked Barker, who could be heard complaining that Suitor was a "goddamned lard ass." (The magazine noted that Barker denied making this comment.)

Just as he had with Nelson Tyler almost fifteen years earlier, Suitor helped Barker get the RB-2000 into working order. The throttle valve was the main issue—it was hard to control and allowed too much fuel to pass too quickly, thus flooding the engine. There is speculation about how exactly this problem was resolved. Barker has said that Suitor, who had been refurbishing the Bell belt at the Niagara Aerospace Museum, took the valve out of the original belt so that he could copy it for his own machine. Some rocket-belt aficionados think Barker simply installed the Bell valve in the RB-2000.

Conspiracy theories aside, after a handful of tethered tests behind the Car Audio Plus building, and a couple of free flights near the airport, Barker deemed the belt ready for prime time. Suitor wasn't so sure. During one of the airport tests, the pilot had had a rough landing, rolling onto his back and dinging the merchandise. It may have looked good— its frame was cherry red, and the silver fuel tanks shone so brightly you could comb your hair in their reflection; Suitor took to calling it *Pretty Bird*—but compared with the Bell and Tyler belts, the RB-2000 was crude and difficult to maneuver. "It was more like flying a truck with a steering problem" is how Suitor described it to Paul Brown.

Nonetheless, when the Houston Rockets swept the Orlando Magic to win the 1995 NBA Championship and were looking to party, Barker booked the gig. At a bash put on by Mayor Bob Lanier, Bill Suitor took off from a barge in the Houston ship channel and made a perfect loop out over the water. Barker's business, now renamed American Flying Belt, was paid ten thousand dollars for the brief air display; Suitor was given twenty-five hundred for his considerable trouble.

In some ways it was an odd choice of locale. That section of downtown is considered less than gloriously attractive. In a short story by Houston native and author Rick Bass, one character describes the Houston ship channel this way: "Here the air was dense with the odor of burning plastic, vaporous benzenes and toluenes adhering to the palate with every breath, and the night-fog sky glowed with blue, pink, orange flickers from the flares of waste gas jetting from a thousand smokestacks."

Toluenes or not, those attending the Rockets' celebration had a blast—Barker's belt was a hit; he seemed poised to follow in Kinnie Gibson's steamed wake. In fact, for various reasons, as we shall see, that is the last time the RB-2000 has been seen in public.

After landing, Suitor gave the belt to Barker, who threw it in a trailer and sped away. "That was the last time I saw Barker or *Pretty Bird*," Suitor told the *Buffalo News*. He returned to Youngstown, his wife, seven kids, and duck decoys.

Larry Stanley, meanwhile, had finally recovered from his head wounds and other ailments. He had only one thing on his mind, and that was revenge. In the fall of 1995 he received a rare bit of good news. The assault charges had been dropped. Adding to his joy was the fact that Barker had been convicted and sentenced to a year in jail (it was later reduced to six months) and eighty hours of community service.

This was a start, but Stanley wanted more. Most of all, he wanted the belt back. He'd caught a news clip of Suitor's ship-channel escapade and, like everyone other than Barker, hadn't laid eyes on the RB-2000 since. So he did what any obsessed, blind-with-rage citizen might do and filed a ten million–dollar lawsuit against Barker and Wright, claiming their lien had been fraudulent. He wanted everything that Barker had removed from Car Audio Plus—especially the RB-2000.

The lawsuit forced Joe Wright into the middle of the Stanley-Barker feud. It was not a comfortable place to be, particularly since, for several years, Wright had been in financial and emotional straits. He'd lost thousands on the belt, the car-stereo business had bottomed out, he was late on his house payments, and his addiction to crystal meth had escalated. He whiled away days futzing around online—playing games, sending e-mails. In early 1996, Car Audio Plus was shuttered, and Wright declared bankruptcy.

And so, broke and desperate, Wright dreaded Stanley's suit and was willing to do almost anything to avoid it, including turning on his old friend and partner, Brad Barker. He told Stanley he'd help him track down the RB-2000 if Stanley dropped the charges against him, and a deal was struck. As part of it, according to *GQ*, Stanley offered to give Wright ten thousand dollars to leave Houston because "[Stanley] believed Barker would kill him for telling the truth." There was good reason to think there might be something to Stanley's instinct. According to Houston detectives, Barker had been making threatening phone calls to Wright, and, as the men's magazine noted, "One friend [of Wright's] says Barker called on Monday, July 13, and threatened to kill Wright if he testified."

Two days later, on July 12, 1998, Joe Wright was due at a six o'clock meeting at the office of his lawyer, Ronald Bass, to complete the deal with Stanley and his attorney, Michael Von Blon. He never made it. Feeling queasy, he asked if he could participate via speakerphone. By late that night, both men and their attorneys were happy with the arrangement.

Sometime over the next forty-eight hours there was a knock at the door of Joe Wright's ranch-style brick home in a Houston suburb.

Wright, shattered, paranoid, and depressed, rarely went out, but he was dressed for clubbing that night in Girbaud jeans, Doc Martens, and a brightly patterned shirt. He opened the door and was mauled by an attacker who bludgeoned him twice in the head with a heavy blunt object. Wright staggered toward the nine-millimeter pistol in his bedroom, but the killer was on top of him, delivering enough blows to render his face unrecognizable; it would take the Harris County Sheriff's Department two days to ID Wright by his dental records. "Joe was unidentifiable as a man or a woman from the waist up," his sister, Nancy, told *GQ*.

Brad Barker was one obvious suspect, but he was not the only one. After doing his time, Barker began shuttling between Houston and Fort Smith, Arkansas, where he was working with an oil-equipment tycoon named Vinson Williams, developing what Barker hoped would be an updated version of the Hiller Flying Platform. The commercial twist to the new machine would come in the form of a soda-can-shaped slot atop the jet-propelled platform where companies like Coke and Pepsi could advertise their wares. "The idea was to start with one," Tom Wade, an aeronautics enthusiast who worked on the project, once said. "Then make five more and have a six pack."

It never happened. The business arrangement deteriorated around the time Wade began to suspect that Barker was pawning parts for cash to live on. Though Williams had already pumped an estimated four hundred thousand dollars into the platform, he and Wade pulled the plug and changed the locks on the warehouse storing what there was of the half-built device.

Soon after, Harris County detectives brought Barker in for questioning about Joe Wright's murder. They interrogated him for three days, hoping he'd crack, but in the end they could not build a case against Wright's former partner and friend. Barker insisted he was in Fort Smith when Wright was killed—and despite his Arkansas alibis' claims that they had not seen Barker on the night of the murder, he was released without being charged.

At least one man remained convinced that the cops had their man. "I am absolutely convinced that Bradley Wayne Barker murdered Joe Wright," Larry Stanley said in an affidavit. "With malice aforethought, by beating him to death in a furious rage with a baseball bat or other blunt instrument."

And although Joe Wright's sister, Nancy, stops short of insisting that Barker is the killer, she does have a long list of complaints about the handling of clues, leads, and, indeed, the entire crime scene. Nancy has said that the sheriff's department ignored possibly helpful physical evidence such as a bloodied bath mat and Joe's pills—amphetamines— that they left behind. His computer was left alone as well—a curious oversight, considering Joe was a notorious and meticulous chronicler of everything and that, as Paul Brown noted, "his entire life was on his hard drive."

Though not discounting the potential of a gay hate-crime scenario, the investigation focused on two local bookies, Bob Malloy and Jeff Warren. Though Joe Wright didn't gamble, he and Malloy had a lengthy and complicated friendship built over fifteen years on trust, secrets, and, as is so often the case, cash.

Malloy, an ex-con, owned the strip club next door to Wright's car-stereo concern. Once, when Wright's business was struggling, Malloy bailed him out with a thirty thousand–dollar loan. To pay him back, Wright arranged for Malloy's daughter to be the beneficiary on a fifty thousand–dollar life-insurance policy. According to Paul Brown, the friends had squabbled about two weeks before Wright's murder when Joe asked Malloy for an additional loan. But when the bookie was brought in for questioning, he passed a polygraph test, and his daughter was suddenly fifty thousand dollars richer.

Jeff Warren worked for Malloy. Years earlier he'd been busted on charges of illegal drug and firearm possession. He and Wright traded frequent personal loans, which had a way of leading to loud disagreements over payment schedules. Warren was also the man who sold

Wright crystal meth. Despite this small mountain of circumstantial evidence, Warren has never been charged with Wright's murder, though he did remain a suspect for many years following the killing.

For Brad Barker, meanwhile, 1999 was shaping up to be a real doozy of a year. In July a Houston judge ruled in favor of Larry Stanley's ten million–dollar lawsuit against Barker and Wright—Barker was legally obligated to return everything he'd taken while Stanley was recovering at home, including the RB-2000. The only problem was, Barker had no way of paying Stanley his money and claimed to not know where the belt was. "Even if I had it," he told Paul Brown, "I would smash it into a million fucking pieces with a road grader." This, of course, did not sit well with Stanley, who allegedly once said to Nancy Wright, "If the authorities don't get Brad, I will. I have friends who will help me. We'll kidnap him, go to the desert, torture him, kill him. They'll never find his bones."

In time, his threat would prove alarmingly foreboding. For the moment, though, Stanley's behavior drifted ever more menacingly into the "arena of the unwell," as movie director Bruce Robinson has described insanity. Once, showing off his newly purchased .40-caliber Desert Eagle pistol in his backyard, he nearly pulled a William S. Burroughs on Nancy Wright when the gun accidentally went off. "I felt the heat of the bullet go by my ear," Wright later said. "It took two fistfuls of sod out of his yard."

Two months after the lawsuit ruling went against him, Brad Barker was standing outside of the Williams Tool Company warehouse in Fort Smith. It was the middle of the night—only Tom Wade and a pack of crickets kept Barker company. Wade stood lookout while Barker shimmied into the ventilation ducts. He'd be damned if he was going to let his latest creation, that soda-pimping flying platform, collect dust. But the instant he wriggled free of the ducts and dropped down into the warehouse, up came the lights, plenty bright enough for Barker to see eight assault rifles staring back at him. Barker had been set up; for the next two and a half months his new home would be a Fort Smith jail cell, where he landed on commercial burglary charges.

And then things turned strange. When Barker got out of prison, he discovered that his Fort Smith–area storage facility had been trashed, turned upside down by an intruder who had made off with anything of value. But that wasn't the worst news. Larry Stanley had sent Barker's mother an extortion letter, explaining that he would return Barker's possessions if Brad coughed up the rocket belt. Stanley told an Associated Press reporter, "I will recover that belt. I'm not anywhere close to giving up on it. [Barker's] going to have to bury it." Stanley even offered a ten thousand–dollar reward to anyone who could bring him his beloved RB-2000.

With his world closing ever tighter around him, Barker was growing desperate for a break. And that's when fortune seemed to finally smile on him. Out of the blue, a Hollywood stuntman whose career peaked with the 1992 Drew Barrymore soft-core thriller *Poison Ivy* called with promises of work.

Chris Wentzel was built like a bowling pin and was apparently almost as charismatic as one, too. He knew Barker through a mutual friend, former stuntman and current rocket man Kinnie Gibson. During a phone conversation in November 1999, Wentzel told Barker he had a three-day gig for him in the Mojave Desert; the pay would be close to fifteen hundred dollars. Despite his lack of experience in the movie business and the curious timing of the call, Barker was in need of cash and so he took the job. He left a message for his Arkansas bail-bond company and was soon on a plane headed for Los Angeles.

Wentzel met Barker at LAX. They stopped for a quick bite and then, oddly, took a short ride in the bay on Wentzel's motorboat before heading for the stuntman's bungalow in North Hollywood. At Wentzel's pad Barker was introduced to two men who would be working with him on the film. The foursome hit it off and chatted amiably for several minutes in Wentzel's kitchen. But then everything went dark. According to Barker's later court testimony, one of the guys he'd just met grabbed him in a headlock, while Wentzel whipped out a pistol and aimed it directly between Barker's eyes. "I just kind of looked at him and grinned because I didn't think it was real," Barker later said in court testimony. "[Wentzel] screamed, 'Get that son of a bitch on the ground.'"

Confused, his mind reeling, Barker was handcuffed, his legs bound by thick rope so that he was hog-tied. "He put the gun to the back of my head and said, 'Last time I put a gun to somebody's head and they smiled at me, I knew they were either crazy or they just didn't give a shit.'"

A moment later, Barker's confusion cleared and was replaced by dread.

"Where's the fucking rocket belt?" Wentzel yelled at Barker. So Larry Stanley was behind this, the captive thought. For the rest of the afternoon Wentzel battered Barker with two questions, "Did you kill Joe Wright?" and "Where's the rocket belt?" If he did and if he knew, Barker wasn't saying. As the sun faded, fed up with the stonewalling, Wentzel ordered a velvet hood duct-taped over Barker's head, dragged him on his stomach into another area of the house, and threw him in a three-by-four-foot wooden crate marked "Scuba Tanks."

"Before he put me in the box, he asked me if I was afraid of rats and snakes," Barker testified later. Then Barker heard a most unfortunate noise: *zzzttt, zzzttt, zzzttt*, the sound of Wentzel power-drilling the crate closed.

The ensuing eight days and nights followed a tight script. In the morning, Barker was dragged from the crate, fed just enough to survive (he told me he had only a little soup and water), let out so infrequently to use the bathroom he occasionally urinated on himself, and questioned over and over on the location of the machine Bill Suitor once called *Pretty Bird*. Some days, Wentzel threatened to kill Barker. On others, he brought up Barker's son's safety as leverage. Nothing.

Losing patience, Wentzel cranked up the torture. One day he pulled the scuba-gear crate into the garage and forced Barker back into it. After introducing that day's assistant as "Jim," Wentzel began drilling multiple holes in the box. "Watch your fucking head," he told Barker. A moment later Wentzel asked, "Do you think that's enough holes?" He answered his own question. "The more holes, the faster it will sink."

With this, Barker broke down. "Please," he begged through his hood. "Please let me out. Let me go. I don't want to drown. Would you do me a favor and put a bullet in my head?"

"Sure. No problem," Wentzel replied. Barker asked for a minute, to say a prayer. He began to pray, but, according to Barker's court testimony, Wentzel interrupted, "If I'm going to shoot, then you're going to be looking at me."

So Barker looked at him—Wentzel raised the gun, aimed, and squirted his victim in the face with a water pistol. "He just kept doing it and was laughing hysterically," Barker testified. It was then that Barker cracked, telling Wentzel that a Texas friend named Steve Mims was holding the RB-2000 in his garage, "not as collateral" for the five thousand dollars Mims once loaned Barker, but just to keep it safely hidden.

On day eight of his captivity, Barker was let out of the crate, unhooded, and yanked into the kitchen, a strip of duct tape stuck like a gnarled silver finger to the side of his head. He was handcuffed to a chair. He was introduced to a friend of Wentzel's, a woman named Elyse Hoyt, who worked as a notary. A small stack of documents sat on the kitchen table.

With his by now well-established distaste for niceties, Wentzel once again flashed his gun, instructing Barker, "You are going to sign, or else I'm going to shoot you in the head."

Barker signed. Hoyt did her part and left the room. "Now," Wentzel announced with obvious glee, "I think it's time you met Jim."

In walked Larry Stanley with his Desert Eagle and down went Barker's jaw. Stanley put the pistol on the table and took a seat in front of Barker. "Where's the rocket belt?"

Nothing.

"If you don't cooperate, things could get worse for you."

Nothing.

Stanley grabbed the freshly inked paperwork and left. Barker was thrown back in the crate. He was stunned, depressed, broken, and twenty-three pounds lighter than when he'd gotten on the plane to fly west. But he wasn't dead, yet. Over the previous couple of days, Barker had noticed that if he wriggled his wrists enough, he could loosen a few of the notches on his cuffs. And then a few more. The day after the Stanley sighting, Barker successfully completed a dry escape run, breaking free of his constraints and trying a window, only to find it stuck. But now he knew he could bide his time, waiting for the right moment to bolt.

The next day, Wentzel left Barker out of the scuba crate while he went on a short errand. This was Barker's best chance, and for all he knew it was his last one. He jiggled the cuffs loose, ripped off the hood, and untied the rope around his legs. Massively dehydrated and unsteady on his feet, Barker struggled to get the window in his room open, but it finally gave. He flung himself outside and ran for freedom.

He ran through Hollywood and kept on running, the cuffs still rattling from his right wrist. He ran for two miles, found a gas station pay phone, and called the FBI. His wrists resembled ground chuck, hacked to bits by the handcuffs. He had last showered in Houston and smelled like a sewer. "I was one nasty son of a bitch" is how Barker put it to *GQ*. The feds met Barker at a nearby diner.

The case against Stanley and Wentzel was not difficult to amass; police uncovered reams of e-mails detailing the kidnapping sent between the two men. Stanley and Wentzel were charged with kidnapping for ransom, false imprisonment, and extortion. Almost three years after they'd lured Barker out west, dangling a Hollywood dream, the defendants stood in Van Nuys Superior Court on the morning of November 22, 2002. Wentzel had taken a plea bargain, and the kidnapping charge against him had been dropped. He was sentenced to seven years in a state penitentiary.

Stanley was not so lucky. He'd refused to cooperate, had denied his guilt at every step. After the jury handed down a guilty verdict in the spring, Stanley fired his lawyer. He was paunchy, graying, defeated, and washed up at fifty-seven. He must have expected bad news, but nothing in the decade that had passed since the American Rocketbelt Corporation had formed could prepare him for what happened next, when in the packed, hushed courtroom, Larry Stanley was sentenced to life plus ten years in prison; he would never be free again. The burly man wept, tears dropping onto his prison-issued blue overalls. "Your honor, I never imagined that I ever did anything wrong," Stanley pleaded with the court. "My search for the rocket belt has cost me more than half a million dollars and left my family destitute and on food stamps."

The judge was unmoved. Stanley was taken to a medium-security jail in the central California desert. But less than forty-eight hours later he began composing a heartfelt letter to the deputy district attorney, Peter Korn, and for the first time admitted his guilt. This was all Korn wanted to hear—he relayed news of Stanley's letter to Judge Barry Taylor, and together they agreed to reduce the sentence to eight years. Brad Barker took credit for the reduction. He told Paul Brown, "I called Peter Korn and asked him to drop the kidnapping charge. I said, 'I'm not asking you to do this for Larry Stanley, I'm asking you to do this for his two children.'"

Whether Barker was telling the truth or not, Stanley's reduced sentence is the last act so far in a twisted tale. It is a tale of passion lurching into obsession, a love of the air crashing into greed, the promise of technology exploding into a nightmare. It left everyone involved shell-shocked. "This is like a Harold Robbins novel," Joe Wright's friend and attorney, Ronald Bass, said at the time to the *Houston Chronicle*. "This is an incredible story of a bunch of very, very unusual people with a unique thing in the middle—this rocket belt." Los Angeles prosecutor Korn told a reporter, "I've never seen people go to such lengths for a contraption so completely outdated."

Which is exactly what interested me. Why go to such lengths? What was it about the RB-2000 that drove men to such depths of madness? How did a fairly useless device hold such power over their imaginations? Or was this even really about the rocket belt? A strange notion occurred to me. I began to think that if I could somehow find the RB-2000, that perhaps buried with it would be a key to understanding the mysterious hold the machine has over certain aspiring rocket men. I began to think of the machine as the Dead Sea scrolls of jetpacks. But where to begin finding it? Of the three partners, one had been killed, one was in jail, and one—well, God knows where Brad Barker is, and even if I could find him, would I really want to get mixed up in his always dangerous, often bloody ways?

Then I remembered something that Stuart Ross had told me when I visited him in the English countryside. It was just as he was letting me off at the train station after our too brief visit on his farm. With a slightly mad twinkle in his eye, Ross had told me he'd heard the RB-2000 was packed up in two waterproof crates and buried at the bottom of the Houston ship channel. I hadn't thought much of this at the time, just the absentminded rumor passing of another rocket-belt obsessive with too much time on his hands. But now, when I thought about it, I had to admit that it made sense—the last time the RB-2000 had been seen in public, after all, was just after its one and only professional exhibition during the Houston Rockets' championship party in 1995. And where did Bill Suitor fly the belt that day? In a loop around the ship channel.

My wife talked me out of hiring a diver to dredge the harbor. Short of that, though, what were my options? It was clear there were only two. I had to reach either Larry Stanley or Brad Barker. And so that's how on a December morning in 2006, while walking Oona to preschool, I came to mail a letter to California inmate No. T75865, requesting an interview.

I figured I'd start with Stanley—at least on paper he appeared to be less prone to violent and unpredictable outbursts. As the months passed without a reply, however, I began to consider contacting Barker. I have to admit, I was nervous about the idea. From everything I'd read, his friends, family, and even associates had a funny way of coming into misfortune.

So I conduct some due diligence, calling a couple of reporters who've worked on stories about the RB-2000 fiasco. One guy returns my call quickly but warns, "Be careful. Everyone involved with this—these people are really, really crazy. Proceed with extreme trepidation." This does not instill me with courage.

Nonetheless, I dial Nancy Wright's number that she'd given me at the convention last fall. At first she sounds suspicious and then just distracted. But she stays on the line long enough to tell me that she still considers Barker a suspect in her brother's murder case. Nancy tells me that Barker has, in fact, been in touch with her recently to tell her that the FBI has finally and officially cleared him, but she doubts this is true. "I believe Brad has a personality disorder," Nancy says to me. I ask what she means. "Well, there's got to be some narcissism in there—it's always all about him. When reality is right there in his face, he has the ability to deny it. He's arrogant enough to think he has the ability to persuade anyone of anything." She tells me he's now in North Carolina, "working a real job because he ran out of people to scam. He's probably working construction, and he'll tell you he's the foreman but I can't see him being a foreman. He's probably just on the crew." I tell her I plan to call Barker, and if I find out anything, I'll let her know. Before we hang up, she says hopefully, "Maybe we can team up to get to the bottom of this."

For many days I don't call Brad Barker. At first, I convince myself this is a situation similar to that in high school where, if you don't tell the girl you have a crush on that you have a crush, she can't reject you. That is, until I hear Brad tell me that the RB-2000 is not at the bottom of the Houston ship channel, for all I know that's exactly where it is. But eventually, I have to be honest with myself. I don't call Brad for many days because I'm scared. Although he may or may not have had anything to do with the death of Joe Wright, he's admitted to taking an auto-body hammer to Larry Stanley and really, really fucking him up. He also had someone beat the crap out of that former Navy SEAL. His enduring rage is certain.

But then, flipping through court documents, I notice something. Brad's birthday, June 14, is coming up—it's less than a week away. This is the motivation I need; I commit to calling him on June 14. I mean, how scary can a guy be on his fifty-third birthday? Still, just to be safe, I buy a calling card and walk downstairs to a pay phone on the street.

After three rings, he picks up—charming, relatively unguarded, twangy. He's back in North Houston, doing yard work. He asks if I can call back later, so he can get on a land line, and we make a plan to talk in a couple of hours.

When I call back he tells me he's now sitting with "one of my attorneys," Jeff Haynes. Then Brad says, "I don't make it a practice to answer calls from unlisted numbers—I mean, I don't know you from Adam—so if you want to call me back from a listed number, we can talk then. I'm not greedy, but the person that treats me right is going to get the greatest story ever."

I tell him I understand and am prepared to hang up, but he hands the phone to Jeff, his lawyer, who tells me that Brad is the one and only person responsible for building the RB-2000, that he is, in fact, responsible for everything except piloting the thing.

I hadn't asked.

"Once you get some in-depth knowledge, you're going to want to meet face-to-face," Jeff says. He tells me the FBI has cleared his client of the murder of Joe Wright and that now they are focused on the guy Brad said it was when they originally brought him in for questioning, Jeff Warren. "Brad is a hothead," his attorney concludes. "But he's not a murderer."

Soon Brad is back on the line, listing, unsolicited, his current impressions of some former friends and associates—Larry Stanley ("not one of my favorite people"), Bill Suitor ("a whiny bitch"), and Stan Casad, who hooked Brad up with Chris Wentzel for that fateful L.A. adventure ("a piece of shit"). It's been nearly a half hour since he told me to call back from a listed number, and we are still talking. In a few hours, Brad is having people over to celebrate his birthday, but for now he is more than happy to ramble on with a stranger about how he is really the genius who built the RB-2000 and about how innocent he is in the Wright case, even recounting several conversations he had with his ex-wife and business partners the day he heard about the murder in his office in Fort Smith, Arkansas, "a fifteen-hour drive from Houston." (For what it's worth, Google Maps puts the drive at closer to eight hours.)

Sufficiently warmed up, I interrupt Barker's monologue for a moment. "So I want to run something by you, a rumor I've heard—I think you'll get a kick out of it."

"Shoot," he says.

"I've heard that the RB-2000 is being kept in two waterproof crates at the bottom of the Houston ship channel."

Barker lets out a tinny, machine-gun cackle. "Yeah, I've heard that rumor, too. And let me just say, I am certain that it is not at the bottom of the Houston ship channel. Look, if I wanted to store something underwater, it would not be in the ship channel—big boats go through there; they'd drag everything and their dog out to sea."

"So," I say, playing perhaps my last card on the matter, "if you are certain it is not in the ship channel, where would you guess it is?"

Another rat-a-tat burst of laughter. After composing himself, Barker sums up the situation, as far as he's concerned: "There will be a movie about all this—no question about it. Let's put it this way," the charming ex-con says, pausing dramatically, perfectly. "If you want to end your book right, come and see me."

A school bus rumbles by the pay phone. My mind flashes to me sitting across the table from Brad Barker, graying but still well coifed, in some sweaty Houston coffee shop, the waitresses fanning themselves with laminated menus. Does this mean he'll tell me where the RB-2000 is? And everything he knows about the Joe Wright murder? The truth?

I tell him I'll think about it. We make a plan to talk again and hang up.

The one thing Barker told me that I buy for sure is the bit about the movie. I don't know if he was aware of it, but by the time I spoke with Barker, Paul Brown had already announced that a film company had bought the rights to his book. And later I learned that a movie called *Pretty Bird*—borrowing Bill Suitor's name for the RB-2000—was to begin shooting soon in New Jersey. *Pretty Bird* stars Paul Giamatti as a local loser who attempts to build a rocket belt in order to become famous. But when he actually constructs the thing, everything goes to pot between him and his colleagues on the project. Sound familiar? I couldn't help but wonder how Brad Barker would feel about the very talented but also schlumpy and chronically disheveled Giamatti essentially playing him on the big screen. Perhaps that was reason enough to head for Houston. But probably not. I'd become more and more certain that Brad Barker would never tell where *Pretty Bird* was hiding.

CHAPTER
7

The Voluptuous Panic,
Part II

With your feet in the air and your head on the ground
Try this trick and spin it, yeah
Your head will collapse
But there's nothing in it
And you'll ask yourself: Where is my mind?

—Black Francis, "Where Is My Mind?"

A week after we get back from L.A., I board yet another 757, this time bound for Europe by myself. I'm on my way to check in with Stuart Ross—he of the near-fatal throttle-valve malfunction—at his country farmhouse in Sussex, an hour's train ride south of London. I'd like to see how Ross is proceeding and also to pick his formidable brain about where I might find a damn jetpack already. At the convention he struck me as a kind and generous chap who wouldn't mind tossing back a few pints and helping a fellow out.

From there, I'll hop across to Dublin to see Will Breaden-Madden. In Niagara, Will had hinted that by this time he might be in the final stages of production on his ShamRocket. And though he is but nineteen years

old and can be agonizingly slow replying to my e-mails with progress reports and a tad flaky-seeming, at this point he remains my best shot at
the wondrously realized future so many others have only dreamed about.

Within minutes of touching down at Heathrow, I'm reminded how
much I love England and the English. Everyone is just so bloody . . .
English. Brilliant observation, I know, but really they could all be telling
me to piss off, and it would still sound completely charming. On the
Tube ride to my hotel in Piccadilly Square—chosen entirely on cheapness and centrality but mostly cheapness, a decision I'd soon come to
regret—I notice a sign reading, "Last Stop Cockfosters." Wonderful. I
leave the station, weighed down by a bulging backpack and a stub-
bornly stiff-wheeled suitcase, and step into a lowbrow comedy about
a tourist in London. I wander in loops, arriving at the same quaint corners over and over again; I look the wrong way and—voosh!—am repeatedly almost leveled by boxy black taxis. I ask for directions,
putting on a vague British accent, thinking it will help, but invariably
approach Russians.

Finally, I arrive at the deceptively majestic entrance to the Regent
Palace Hotel. Don't look for it—mercifully, it's no longer there. From
what I understand, soon after my stay it was converted into shops and
offices. Which is just as well because get past the soaring, Victorian
facade, topped by the bright neon wings of—who, Nike?—and the
decent-enough lobby, and you enter hell on seventy pounds a night.
Only smellier. All the Lysol in the world cannot conceal the many years
the Regent has waged a losing war against the publicly urinating, the
chain smoking, and the vomit prone. It is a foul, foul stench that greets
me as I step off the lift on the eighth floor and navigate a warren of
wounded halls, the destroyed dingy green carpet held together in an
alarmingly high number of places by black duct tape. From what I can
tell, there is but one elevator bank per floor, which, when combined
with the hotel's labyrinth-gulag design motif, means it takes a solid
ten minutes to walk to the door of my room at a brisk pace.

But just beyond the communal bathrooms, inside of which I can see the graffiti of either a monster or a dragon fellating itself and the words "Late Night Lovin'" in thick marker, I land at room number 8075. It's modest—bed, sink, mirror, telly. A delicate drawing of the *Hemerocallis valentina,* which could be the offspring of an onion and a daffodil and named for the porn industry, hangs on the wall. To shower, pee, or worse, one must brave the battered hall, forever crammed with drunken teens. Drunken teens, actually, could very well be the hotel's best feature.

But no matter—I'm on a mission. And no puke-smelling underage playground of a tourist-trap hotel is going to slow me down. The time difference, however, will. It's now nearly eleven locally, too late to call Ross. I'll ring him up first thing in the morning.

I go in search of food and pass up a homey-looking pub serving fish and chips for a falafel chain. The narrow Piccadilly streets are bustling with preholiday throngs, who clatter down brick alleys veining off mighty Oxford Street, their faces flushed and glowing in the ember of cigarettes and the faint buzz of Christmas lights, bizarrely featuring a cartoon frog where Santa Claus should be. I wander suddenly, unexpectedly, into the red-light district, where one club proclaims with typical English modesty, "Possibly the most exciting gentlemen's club in the world." A few more blocks, past darkened West End theaters, dodging homicidal taxis, and then, exhausted, back to room 8075.

In the morning, I call Ross from the lobby phone but get the machine. It is only now I remember that Ross had written in an earlier e-mail that he and his wife are expecting a baby right around this time. And then, before I can help myself, I'm actually thinking, "Don't fuck this up for me, unborn kid!"

I'm alone in a land that is as close to being New York City as just about anyplace else on earth, yet I feel utterly foreign. I leave messages for a couple of friends of friends, hoping for some sort of human contact today in case Ross is tied up at the hospital—a possibility that is seeming more and more likely as the minutes evaporate. I'm due to fly to Dublin in three days and so already panicking slightly about making the Ross meeting happen at all.

Unsure what to do, I wander. Piccadilly Circus, unfortunately, looks no better in the daylight than it did last night—schlocky gift shops, bad restaurants, lonely-looking tourists. I don't want to stray too far as I need to be close enough to grab my bag and bolt out to the country as soon as I make contact with my suburban rocket man. After lunch I leave another message for Ross. I don't want to disturb the guy if he really has just experienced the glorious miracle of birth, but still. Over the past few months I've grown increasingly desperate to locate my jetpack, and so every tick of my cheap watch is like another tiny dagger to the heart.

In the evening, I meet an editor friend for a couple of Guinnesses at a nearby pub, then rush back to the lobby phone bank. I've now been in London nearly twenty-four hours and am no closer to connecting with Ross. What if it isn't the kid but something terrible and tragic? What if there was an accident? I dial, get the machine. Jesus.

Exhausted and more than a little shaken now, I sit on the spongy, faux-leather bench beneath the pay phone. On the other side of the wall, behind my head, I can hear the *giddyup, up, up, thumpa, thumpa, thumpa* of a truly awful ska band incongruously playing a *Titanic*-themed party in the hotel's banquet room. Every now and then someone walks by in a historically accurate costume. I see men in tails and monocles, women with abundant plumage.

I must not be hiding my creeping dismay very well because a party waitress in a white oxford shirt and a nose ring leans over and asks, "Are you OK? You look sad." I start to explain my ridiculous circumstances—over here looking for a jetpack, the guy may have just had a kid—when, like magic, or the movies, the phone rings. I look at the waitress. "You'd better get it," she says. I pick up the receiver tentatively. It's Stuart Ross—he's been at the hospital all day. He sounds tired but thrilled. "I've got me a little rocket man!"

Unfortunately, Ross can't meet for a couple of days, so after hanging up I wander alone again. All the decent-looking restaurants are already closed. Feeling defeated, I break down and buy a Whopper with cheese from the Burger King across the street from the Regent Palace. Back in my room, beneath my *Hemerocallis valentina*, I pathetically gobble the burger. I'm alone. As alone as Juan Lozano was for forty years, tinkering away on his many rocket belts. As alone as Jeremy McGrane in his parents' garage, a welding mask lining his face. And Gerard Martowlis in Jersey, sneaking to the basement for a fix after his girls are safely tucked in for the night. Or even Bill Suitor, twelve hundred flights under his rocket belt, and each one of them a solitary mission.

Two days later I'm at the Victoria Station way too early in the morning. Ross has carved a couple of hours out of the postpartum insanity that is now his life. The train pulls out of London, and I can see the raspberry-brick buildings below the tracks. Right on cue, lyrics pop into my head. "On the rooftops of London. . . ." *Mary Poppins*, Oona's current obsession, is yet another way flying fantasies get us when we're young.

We zip past verdant fields holding empty soccer goals, and an hour later I'm outside of the Horsham station in Sussex, quite likely the quaintest old train depot ever built. While I'm deep in a reverie concerning the lovely, quiet countryside, the many uses of brick, and *Masterpiece Theater*, Stuart Ross rolls up in his silver Mini Cooper, '80s pop music rattling synthily from the speakers. I get in, exchange pleasantries, and we tear off toward the farm.

Ross turns the music down, which is good because we are seriously booking along tight roadways, cramped further by two walls of shrubs sprouting at the edges of the pavement. We blast past bicyclists, with mere inches to spare. But Ross—youthful despite his craggy visage and graying cropped wedge of hair, a master of the 757 with more than twelve thousand hours in the sky—is in command.

We squeal into the driveway, a wide slab tucked between the main quarters, a hundred-year-old farmhouse Ross has renovated, and the barnlike garage where he assembled his rocket belt. The house sits on an emerald acre. I'm reminded of something Nelson Tyler told me back in L.A.—to really pursue this dream, a man needs three things: time, space, and money. Ross would appear to have nailed the trifecta.

As we climb awkwardly out of the Mini—and, really, if there's any other way to exit the thing, please do let me know—I spy the pirate's plank where Ross almost killed himself in the distance. I'll have to take a closer look, but for now we gather around the cozy kitchen table, me on a church pew that's been appropriated as a breakfast nook.

On the table sits a copy of the venerable old satiric magazine *Private Eye*, Ross's BlackBerry, a Sudoku book, and a pack of Marlboro Lights. As we talk, Ross jumps up every few minutes to light another cigarette off the gas stove, let the dogs out or in, or answer the phone for more congratulations on the birth. The decor is pure domestic-rustic hominess, all dark wood and low ceilings—if skiing lodges existed in England, they might look something like this. The walls are dominated by images of lizards (presumably drawn by Ross's older kids), the Simpsons, and the family's two fluffy golden retrievers, Lilly and Molly.

I'm sort of stunned that three days ago, Ross became a new dad, yet here we are, talking rocket belts, jetpacks, and other lunacy while he waits for his wife and baby to come home from the hospital. I chalk it up as a testament to the British affinity for good manners and the immense and enduring devotion this man has to flight. A devotion, he reminds me, that is more than twenty years old now. As a teenager, Ross caught the flying bug when a colleague of his dad's (the older Ross worked as a Kodak salesman) took father and son for a spin in his plane. "What really got me was all the knobs and buttons—I thought, 'Fantastic!' And that's how it all started, I guess."

By the time he was a young man, Ross was working as a Cessna flying instructor. It was then that he saw Bill Suitor's Olympics act on television. "I just thought it was amazing," he tells me now. "It's one of those things—it's American, it's over the top, it's fantastic." But despite being blown away by the images on the screen of a man soaring "with no wires attached," as Jim McKay put it, Ross didn't give the idea much more thought. This was pre-Internet, after all, so procuring information would have taken more time than Ross could afford. "To be honest with you, life moved on—I started flying commercially."

He first flew wide-bodied 747s, then later 757s to Australia, Singapore, the States. When Ross was promoted to senior pilot, the flights became shorter and less frequent. He began spending more time at home. He fixed up the house, looked after his dogs, mowed the grass.

There was one idea, however, he couldn't shake. About four years ago, with the Internet now ubiquitous, Ross sat down at his computer and typed into a search engine the words *weird flying machines*. Before too long he had tracked down the e-mail address of a NASA guy in the States named Tim Pickens who appeared to have done some rocket-belt tinkering of his own. Ross arranged for Pickens to ship him all the materials he would need to get started building his machine, which was three aluminum tanks, a corset, handlebars, nozzles, an engine inside which the catalyst pack could be placed, and assorted other nuts and bolts. Pickens said he'd also throw in a couple of rocket-belt building books and some handwritten plans that he'd gotten from the Smithsonian Institution. He wouldn't need the gear any longer, as he was going to work with Burt Rutan on the civilian spacecraft SpaceShipOne.

It was then that Ross began to think his real dreams might actually morph into reality. The pilot was already having nightly flying dreams, wherein he'd strap on a rocket belt and zip around his garden. "That is the bizarre thing," Ross says to me now, before popping up from his kitchen chair to stick another Marlboro Light into the stove's gas flame. "You might think I'm crazy, but it is absolutely true."

There would be other foretelling events. Not long before Christmas 2002, Ross was in the kitchen when FedEx delivered Pickens's package. The television was on. It was early evening, the country light softly fading. As the Englishmen stood contemplating the brown box, the opening sequence to a *Simpsons* episode flickered behind him. Even today Ross looks giddy and wide-eyed as he tells me that on that day, the famous show-starting Simpsons antics—the family navigates myriad Springfield traffic hazards and manages, just as the theme music ends, to all arrive if not happily then at least together on the living room sofa in front of their own TV—included a twist. "You know how Homer drives into his garage and they all go run through the living room?" he asks me, his small eyes open wide. "Well, that day they all have jetpacks on, and they're all flying around. Absolutely, absolutely true. And I thought, 'Well, that's an omen.'"

And not the last one, as far as Ross is concerned. He tore into the Pickens box, equally thrilled and overwhelmed by the various gleaming parts. Quickly realizing he was going to need help, he enlisted the services of an engineering friend (Ross himself, outside of his flight school, has no formal training in aerodynamics or aeronautical engineering), and the two men got to work in the garage out back. "I said to Mike, 'Look, there's going to be machining, milling, lathe work, welding, electric—a lot of it will probably be trial and error, so let's just see how we get on with it.' And he goes, 'No problem, Stuart.' He honestly thought I was completely bonkers; he thought he'd never see me again."

Instead, his friend Mike started seeing more of his old chum Stuart than his own family. Every free moment they were at it. The hard work paid off. A year later, largely on intuition, they'd re-created the Bell Rocket Belt. There was only one problem. It didn't fly. "I said, 'We're going to have to scrap this, Mike.'"

And they did, salvaging only the original fuel tanks that Pickens had sent. Undaunted, the two friends hunkered down to construct a second belt. This took another year. "And you know what?" Ross says to me. "The following year, we did the same thing again. A lot of the plumbing wasn't right, the nozzles weren't big enough, the engine—at the time we thought it was too small, and we put a larger engine on, which was a mistake; we should have stuck with the original engine."

At this point, most people would have packed it in. Actually, most people would have packed it in long ago. But one of the more curious things about the machine is that for certain men scattered all over the globe, there is no greater endeavor. And nothing short of death is going to stop them from trying. As the roboticist and author Daniel Wilson once told me, "There's a huge percentage of people who say, 'Where's my jetpack?' And a much smaller percentage who say, 'No, really—where's my jetpack?'"

Ross is clearly and firmly in that smaller percentage. Luckily for him, and with additional guidance from Nelson Tyler, Bill Suitor, local machinists, and others, the third attempt was successful. On April 20, 2005, not by accident forty-four years to the day after Harold Graham blasted off for the first time, Ross clipped a steel cable to the pulley above the test platform and cracked the throttle valve. He could only pray that the pressurizing nitrogen tank an inch behind his head, containing 5,000 psi, would hold, because he knew very well what sort of damage an exploding tank would do to a human skull.

The tanks held. The rest of the equipment also worked beautifully. Ross felt a forceful tug under his outstretched arms. Three years and countless hours after that fateful FedEx delivery, Stuart Ross's feet hovered about ten inches off the ground. Mission accomplished.

Over the spring and into summer, Ross executed similar tests another eleven times, each one an ecstatic, if limited, success. He was anxious to fly free of the tether wire but knew that for his own good he shouldn't rush it. Besides, he was confident he'd eventually be able to soar like Eric Scott does for Go Fast!—and perhaps earn as much as the Powerhouse guys, too. He'd get there one day, he was certain. It was, Ross increasingly believed, his destiny. "I don't mind admitting," Ross tells me in the kitchen, "and you might think I'm a complete crank, and I know Wendell Moore died years ago—but I think there's somebody up there that wants to see this project work. And there's somebody, definitely, that wants to see this thing flying with me on it."

If Moore or some other spirit is playing fairy godfather to Ross's rocket-belt dreams, it certainly has a wicked sense of humor, for it was on his thirteenth tethered test that Ross's hydraulic lock malfunctioned, causing his throttle valve to jam, at which point ensued the plastic-bag-like whipping-around business and the headlong crashes into both the nearby fence and his mate Mike, which of course solved Mike's chronic back problems. "It was horrendous, just absolutely frightening." Ross turns sober, picking at the skin on his hands. "I had a three-eighths-of-an-inch steel wire holding me up. As I was spinning around it like a Catherine wheel, I thought it was going to go around my neck and decapitate me." Ross reflects on what he just said for a moment. "I was getting blasé about it. And I don't mind admitting: we used to do the craziest things. We used to fuel it up in the workshop, then charge it up with gas, and then I'd get on one side, my mate would get on the other, and we'd just walk it out. Put it on its stand, I'd strap it on, and I'd fly. But we don't do that anymore—it's all done outside. So if it starts to leak, we know we're not going to burn the place down. But you don't think of it at the time."

Ross was fortunate to limp away from the disaster with only a severely bruised knee—and ego. The BBC was filming a segment on his project that day, and the supervising producer was eight months pregnant. "I thought she was about to have it." Ross is laughing because all's well that ends well. "She's going, 'Huh! Huh! Huh! Huh!'" His eyes are nearly bulging out of his head. It turns out the noise that's generated from having twice the typical amount of fuel surging out of the machine's nozzles is quite something. "A friend was doing some gardening about half a mile down the road from us, came around the next day, and said, 'Stuart, what on Eeeeaaarrrthhhh were you doing about half past two?' He thought the house next to him was falling down."

Then again, maybe Wendell Moore really is monitoring progress out here on the Sussex farm, because things could clearly have been a lot worse. The thirteenth test was, after all, going to be Ross's final ride on the safety wire before flying free. What if this had occurred on the fourteenth try? The fuel tanks can carry enough liquid power to launch a man eight thousand feet into the sky if he flies full-bore straight up. "With that amount of fuel going through it, it really buggered up the catalyst pack."

Some throttle adjustments were in order. Ross ditched his current valve (the fifth he had made) and tracked down the original Bell blueprints. It is a poorly kept secret that Ky Michaelson will sell, on a sliding scale that ranges from a barter for spare parts to a few thousand bucks, the plans to pretty much any taker. No one seems to know how he got them. "I don't want to see you kill yourself," Michaelson said, according to Ross. "I would never give these plans to another American or a Frenchman, but I love you British guys."

With the throttle plans in hand, Ross seemed to be honing in on a final product. But there were yet more obstacles to come. "We've had some really weird things happen to it. We've had it shaking so violently that you couldn't hold on to it—there was too much silver, and the gas was not reacting and was not able to expand down the tubes. That was when we started running it after the accident—the silver had gotten so hot and had literally welded itself together because we ended up running at such a high fuel level. We looked inside it, and it looked all right. It's called chogging, that's the technical term for it where the gas that is generated, so the nozzles were severely vibrating, I'm talking two or three inches, it sounded like 'brrrpp, brrrpp'—you know the tongs you use in a garage to take the wheels off? I sat down on the stage out there with it on, thinking, 'What the bloody hell is wrong with this thing?' And you know what?"

I didn't.

"We thought it was something to do with the nozzles, that the two tubes weren't big enough, so we replaced them and put new nozzles on. And that cost probably seven or eight thousand bucks to do that, and it took weeks and weeks. And we put it back on, put the new nozzles on, fired it up, and the same thing happened again." He looks completely perplexed, mouth hanging open. "You know?"

Disaster struck again when a shipment of concentrated hydrogen peroxide rocket fuel from the Swedish chemical engineer Erik Bengtsson to Ross sprang a leak. The truck carrying the fuel exploded in flames on the always-busy M25 roadway, not far from Heathrow. As *New Scientist* magazine reported in a story on Ross, "A large section of the road burst into flames, shutting the motorway for hours and causing chaos, though there were no serious injuries."

"It was a complete nightmare," Ross tells me. The accident cost three hundred thousand dollars to clean up. Bengtsson was shaken. The Health and Safety Department was concerned, particularly in light of the fact that hydrogen peroxide is what the UK Tube bombers had used earlier in the summer. In the end, though, Ross was only quite reasonably forced to move his gasoline inside, rather than keeping it out near the garden.

We take a break and stroll outside. I'm anxious to see the garage, where all of this madness and ferocious tinkering has gone on. Ross leads me to the shed and swings the creaky door open. Inside, several small connected rooms are packed with tools, parts, assorted knickknacks, and equipment, not just of the rocket-belt sort. One room is home to a pair of slot machines, the faces removed to reveal the blinking innards, the detritus of a "typical lad," who once fancied taking radios apart and has graduated to slots. In the shadows I see a cigar-shop Indian. Beyond that, if my eyes aren't playing tricks on me, is a full-blown disco with polished dance floor, a mixing board, a mirror ball, crates of CDs (a Michael Jackson collection sits atop one stack), and an elaborate lighting system featuring "more bulbs than they have at the local club," Ross tells me. He and his wife like to entertain. Or did, anyway, before this week's birth of his rocket man, Ben.

Passing the disco room, we are next standing beneath a white Lawrence of Arabia–style tent where much of the rocketing work has been done. Laid out on a card table are several throttle-valve iterations, the attendant pins and clips, and a pair of white sneakers singed brown-gray by a few burning drops of concentrated hydrogen peroxide. Nearby are the accoutrements I've become familiar with as the necessary accessories for anyone fancying himself the next Buck Rogers, namely, a power drill and a welder.

And there, resting on its metal easel, is Ross's masterpiece. It is sometimes remarked on that dogs tend to look like their owners. I might say the same about rocket belts. Or at least I am starting to believe the approach taken by the builder is borne out in the completed design. Where Juan was methodical and painstakingly precise, Ross has been spontaneous in the extreme, even having "no idea" what he was doing for the first couple of years of work. So perhaps it makes sense that Juan's belts are sleekly polished prizes and Ross's belt, before me now, is a gorgeously grizzled mutt. The red paint of the corset and fuel tanks is chipped here and there, no doubt due, at least in part, to those unexpected collisions with fences. Union Jack stickers have been slapped on. The nozzles look made of foil—they aren't, of course, but the contrast to Juan's carefully brushed stainless steel is striking. Ross's belt is tough looking, as though it's been through a war, reminding me of the way *Star Wars* equipment—land speeders, 'droids, blasters— was often packed with the sand and grit of Tatooine. It was of the future, but not an antiseptic one.

And hey, if the thing flies, who cares what it looks like, right? Ross swears we should all know soon enough if his belt can fly for real—off the tether, twenty, thirty feet into the air. He is loath to put a timetable on it, but that day he tells me he is sure he'll soar sometime over the coming winter. But then he laughs. "What I'll say is, I'll definitely be off the tether within five years. Fair enough?"

He's joking, of course, but even if he weren't, I don't doubt that if it means plugging away for another five years, he'd do it. That despite the fact that he estimates he spends fifteen hundred dollars every other month on fuel and anywhere between three hundred dollars and a few thousand every month on various engineering costs. So far, he's in for a total of about four hundred thousand. But he doesn't know for sure because he hasn't kept track. "I don't want to know, and I don't care what it costs, to be honest with you. If it costs twice that at the end of the day. . . ." His sentence drifts off before picking back up someplace else. "It won't do because really I think we're at the end of the development stage."

"Why?" I asked Ross earlier that morning. "Why is it all worth it?"

He didn't pause. "It's the ultimate David Copperfield magic trick—and it is almost magic, really." The voodoo that Nino had spoken of weeks ago in California.

Now Ross wants to demonstrate the power of peroxide. We shuffle outside. Using metal tongs, he picks up a single silver screen, about three inches in diameter, and dunks it into a glass measuring pitcher containing pure, highly concentrated HP. Immediately, a cloud of steam pops into being. It has a sour-sweet smell and brings to mind rock-concert fog machines. There is almost enough of it billowing around our legs to satisfy the first three rows at a Mötley Crüe show. The loud hissing sound could be that of a pissed pack of hammerheads. Ross's catalyst pack contains one hundred of these silver screens.

We wander over by the wooden stage where my host has now completed some thirty-five test runs. It's a brilliantly bright early fall afternoon. There is the faint call of birds in the distance. Occasionally, a car crunches down the gravelly road running past Ross's house. The air is calm, and it seems like the right moment to ease into the other reason for my visit today. "So, I'm wondering, when do you think we'll finally see jetpacks replace cars?"

"It won't happen, just won't happen. Too expensive, too complicated, too difficult to train—it just won't happen, not in our lifetimes." He didn't even need to think about it. Or maybe he already had. Like Juan, Stuart ultimately hopes to start earning his half a mil back someday. And perhaps it is understandable if years and years of toiling and a major outpouring of pounds have altered what for others remains a boyhood Buck Rogers fantasy. "This has only one purpose," he'd told me earlier in the day. "At a big event, a show, and that is it. A thirty-second show, if that."

I ask, "Will you fly your belt at big events someday?"

"Yeah, halftime for football. Somebody opening up a shopping mall or something. A big day, have the Rocketman there. The following is so huge—everybody loves it. You saw when Eric landed [at the convention in Niagara]—everybody . . . woosh! Cameras, interviews, newspapers, TV—they love it. Everybody loves the rocket belt. Even my wife thinks it's fun. As long as I test it when she's at work."

His wife! The day has disappeared, time melted. We've been at it a few hours now, and Ross is due back at the hospital. He has a new future to tend to.

We drive the same whiplash pace back to the station. Ross bounds into the lot and lets me off at the curb. Before he rips away, I lean my head toward the open passenger window of the silver Mini. "Good luck," I say, believing more than ever that luck has nothing to do with it. Then he's gone.

By the time I get back to London and drop my bag at the hotel, it's early evening. Tomorrow I leave for Dublin—to see Will and, hopefully, his

ShamRocket. Tonight is my last chance to see a Tate exhibit that sounds intriguing, so I hop on the Tube, get off near the Thames, get lost, and eventually find the Millennium Bridge that takes one over the river to the museum. As I clack my way across the suspended structure, the sun is in deep retreat; to my right I can see Big Ben glowing in an orange halo.

I'll only have about an hour inside the museum before it closes, but that should be plenty of time. I've come to see the work of a hot-wired German artist named Carsten Holler. In taking over much of the ground-floor real estate of the museum, Holler designed a series of spiraling, brushed-stainless-steel tubes with Plexiglas roofs, meant for sliding. He'd named the series of slides *Valerio II.* The name referred to something Holler had once heard at an Italian rock concert in the summer of 1998, a fan in the crowd pretending to know the name of a sound technician working the show who'd wandered off and was needed. "Valerio!" the fan cried out, as a prank, and soon the entire audience had joined in. "There is something about the sound of this name that makes you want to shout it out loud," Holler told *ArtForum* in 1999. "You feel a little better after you've done it, just like after having traveled down a slide."

The artist also believed the project might yield related but perhaps unexpected dividends. "A slide is a sculptural work with a pragmatic aspect. It can be used as a means of transportation—one that is effective, environmentally sound, and elicits happiness. You let go and lose control, and a moment later you arrive safely at another place."

This sounded familiar. In the vast Tate, Holler had erected five different slides, each beginning on a different floor and funneling riders down to the ground level, where they exited onto gymnastic mats. The longest slide is 160 feet.

This sounded good on paper, but it isn't until I'm standing on the museum's bottom floor that I'm able to appreciate the slides' manifest awesomeness. For the forty minutes that I watch, each of the stretched cylinders contains at least one whipping human missile. Whooping calls of joy and whistling float through the hall. High school girls, New Wave dudes, football moms, all-business dads, locals, tourists, art-school dropouts—it seemed to matter not at all who was sliding; the result was pure euphoria as another shrieking body came thumping to the canvas. "To shout 'Valerio,'" Holler had concluded, "is, of course, desperate and hopeless, but it provides relief from the burden of straightforwardness."

Isn't this what Stuart Ross and Juan Lozano and Jeremy McGrane and Tom Edelstein, Gerard Martowlis, and all those other aspiring air-borne commandos are trying to do—find some relief from the straightforward, the mundane? Even if for only half a minute at best?

I circle the landing mats to get a closer look at the exhibit's accompanying text. There, on the paper plaque, are included the words of French writer Roger Caillois. Sliding, Caillois believed, provides "a voluptuous panic upon an otherwise lucid mind."

A voluptuous panic—it doesn't seem entirely pleasurable, yet I can't deny the surging feeling that this is something I wanted to try. But the mysterious sensation would have to wait another day—the museum is now closing.

I wake up the next morning with a song if not in my heart then at least very near my spleen. Today's the day I go to Ireland, first to Dublin, where I'll make a plan to meet up with Will Breaden-Madden, either at his folks' place in Longford County in southern Ireland or in Belfast, where he's in school. Although I believed, and still do, that it was important to see Juan and Stuart, to appreciate all the work they'd done on their flying machines, I think I've known all along that my best jetpacking hope rests with the scruffy-bearded nineteen-year-old whiz kid. I'm excited to see if my hunch is on the money.

Also, today is the day I'm saying good-bye to the monumentally smelly Regent Palace, and though I normally hate good-byes, this one is most welcome. And so I schlep out to Heathrow, board an Air Lingus jet, clear across the Irish Sea, into the perpetual Dublin drizzle.

I'm staying with my wife Catherine's younger brother's wife's family—got that?—the Stones. I've previously met the mom, Barbara, and dad, Graham, once and each of the kids, too—Lucy, Pinn, and Merlin, in descending order of birth. From what I can tell, this is a true bohemian clan—an ad-writing, gastronomic patriarch, a tough and sweet mom, and three preternaturally curious homeschooled kids. Pinn and Catherine's brother Billy lived with us in Brooklyn for about two years, helping us survive Oona's early years. When I return from this European adventure, Merlin's going to take over for a three-month stint as head nanny.

The house is four stories and not much wider than the wingspan of a Cessna. It is 180 years old. The Stones purchased it in 1981, back when the surrounding neighborhood of Portobello was a rather sketchy place and the thought of a resounding Irish financial boom was laughable. Lucy tells me over tea that she's pretty sure the place was once a brothel, due to the fact that for years after they moved in, shady-looking men would swing by at odd hours asking, "Is Mandy in? No? How about Sheila, then?" The walls are packed tight with books and homebuilt flourishes such as the dining room wheel utilizing piano keys as spokes and a ship window at the center. This, of course, means that Graham actually has reinvented the wheel.

His third-floor office is crammed tight with foodie books, each title more fantastic than the one next to it: *A Freezer for All Seasons*, then *Queer Gear: How to Buy and Cook Exotic Fruits and Vegetables*. (When I arrive Barbara, Graham, and Pinn are in Rome on an eating tour, but Dad has left behind his essence in the form of a large pot of refrigerated, congealed duck fat.) There are also three Macintosh computers, each from a different era, in the study, along with eleven pen-filled mugs lining the windowsill—Varsity Marmalade, a 1979 commemorative pope cup, et al.—and too many work-related cassettes to count.

Where there might be a speck of bare wall peeking out, Graham has solved that problem by nailing up tin-plated signs from his travels that serve as de facto allusions to his bemused worldview. "Loitering & Soliciting in this building prohibited"; "Sorry no credit except by prior arrangement"; "Don't forget to bring home the dogs."

I settle into my top-floor room and call Will. I'm tense with expectations as the low, fuzzy growl of the ring kicks in. And then his dad answers and tells me Will is in Scotland and won't be back for a week. What? No! I'll be gone in a week—there must be some mistake. No, that's what he said, Mr. Breaden-Madden tells me. Oh, Jesus, Joseph, and Mary! This can't be happening—I'd e-mailed Will a week ago to confirm.

I hang up and contemplate my future. I should have known better than to trust a teenage theoretical-physics major building a Sham-Rocket. The irony is, at the moment, more painful than cute. But let's not completely freak out just yet—perhaps there is something I don't know. I pass the night shivering in the Dublin chill, drinking cans of lager, and—I can't help myself—eating fish and chips.

The following afternoon brings a repeat performance of my London days. As was the case with Stuart Ross, just as I'm ready to forget all this craziness and get home early, the Stones' phone rings.

"Uh, yeah, Mac Montandon?"

"Will! Will, where are you?"

"Uh, yeah, that's the ting."

"What's the ting?"

"Well, I think there was a bit of a misunderstanding. . . ."

"I'll say."

"I was in Scotland but came back early to meet up with you—I guess I forgot to tell my dad. I'm in Belfast, now."

I tell him I'll be on the early train the following morning.

Meanwhile, Merlin's throwing his own going-away party tonight, so as early evening approaches I'm huddled around the kitchen's wood-burning stove with my host, his surly but sincere friend Johnny, drying pot in a paper towel on the stove, a pair of giggling twin sisters, and one other slight-looking young woman who handles her beer bottle in such a way that I have little doubt she can drink me under the table.

I need to turn in early, but soon the house is humming with late-teen party machines wearing all manner of questionable hairstyles. I take a seat near the living room fireplace. Beside me is a Dylan doppelgänger, a molecular biology student, and his redheaded girlfriend/Muse. Dylan is rolling a cigarette out of Copenhagen hash on the street map I'll need to walk to the train station in the morning. I make a mental note that I'll probably need to find a different map in the morning. Dylan puts on a mixed CD he made for the party and hands me the cigarette. Already feeling about eighty-seven years old in this crowd, and for reasons that are beyond me, I decide it would be more unwise to beg off partaking than to inhale. A few moments later, I am absolutely certain that hidden in the music of Marvin Gaye is the answer to solving all earthly troubles. It must be time for bed. I say goodnight and stumble upstairs.

Some sights from a northern-bound train: flawless rugby fields, gray-slab housing units, abandoned cars on a demolition derby track, a silent sea warbling up to a string of narrow beaches. Then: sheep, sheep, and more sheep. I literally count them until I've drifted off to sleep. When I wake up, we are there, the train hissing to a halt beside a long wall splattered with Technicolor graffiti and topped with a dollop of barbed wire.

I have to admit, I'm a bit surprised to see that Will is actually waiting for me in the Belfast station. He's wearing the same fitted fatigue jacket, black T-shirt, and quasi–Indiana Jones hat he had on in Niagara. His look. I'm even more surprised to learn that Will has made a lunch reservation at the swanky, ornate Merchant Hotel and that we take a taxi there. My teenage host pays for the cab, reaching into a brown envelope for the paper pounds.

Inside, it's all rose velvet banquettes and chairs, gold-filigreed ceiling, stained-glass mosaics, and piped-in harp solos. The whole thing feels peculiarly romantic—hyperromantic, even. I count more tiny cupids than diners.

As we look over the menu, I ask Will if there is anything he doesn't eat.

"Well, I don't eat monkeys," he says, as if he'd been asked the question too many times already.

The waitress takes our order and when she sets down bowls of cream of parsnip soup says, "Enjoy." As she walks away from our table, Will leans in close and almost whispers, "If anyone says, 'Enjoy,' what they should really say is, 'Enjoy it.' That's correct grammar."

He's right, I suppose, but still.

I turn the conversation to the ShamRocket, which, I couldn't help but notice, is nowhere to be seen. Will tells me it is being stored at his parents' place in Longford County, and there's no way we can go there today, since he has a presentation to give tomorrow on the topic of teleportation. "Of atoms, not apples," he points out. But that's okay, he assures me; he's just a few months from being ready to publicly demonstrate his ShamRocket, and he offers to fly me back over at that time to be the first journalist to cover the event. He's thinking February 13, picking the unlucky number "just to mess with people." He's already begun alerting friends and family and Richard Branson, so that they might save the date.

Then Will gives me his quick history—a lifelong fascination with flight, a boyhood ability to make anything, really, and the burning desire to solve, once and for all, the problem of air-time duration that's plagued wanna-be rocketeers for close to a century. He tells me that his parents—Dad's an architect, Mom a psychologist—have helped pay for his work so far but that he's also secured funding from a European mogul who'd rather remain anonymous. "In case things don't work out, he doesn't want to look bad. That's all, then."

Will reached the unnamed mogul through a letter-writing campaign, for which he drafted notes to some of the world's richest men, Sir Richard Branson among them, seeking financing for the ShamRocket. "I'm pretty good at getting people to do what I want," he concludes.

"Well, I'm here, aren't I?"

We both laugh—me a little uneasily. I push away the thought that I'm being strung along. Really—to what end?

When we first began corresponding months ago, Will had made it sound as if I were mere weeks away from flying his jetpack. He'd written: "The ShamRocket 2 will be able to run on a variety of fuels (including Diesel and Jet-A1) . . . it will have a flight time of 10–15 minutes (this should improve, but is the estimate for my prototype). . . . If you hold off on coming over for, say, a month, I'll probably have the Sham-Rocket 2 completed, then I'll let you try it out." And later he'd written:

> I currently have both engines fully working, but I am continuing to test some aspects of them. I have almost all of the chassis completed, almost all of the electronics completed and just a few minor odds and ends to acquire. I have designed a safety device for the unit also and I am currently researching this. You will notice that all previous rocketbelts have had no real safety/back-up system. I anticipate that the safety system I have designed will be so safe that practically anyone can fly this jetpack in complete security. In other words, the device should not be available exclusively to pilots, but to the general public and may genuinely provoke a revolution in transportation.

Will pulls more bills from his mysterious envelope to pay for lunch. Now we are in another cab, speeding toward the Ulster Flying Club on the city's outskirts. Will wants to take me up in the plane he rents there. We pass Northern Ireland's Parliament building, a dark castle set quite far back from the road at the end of a mile-long driveway, as if, one can only imagine, to forestall any unwelcome guests.

Suddenly, Will pulls a gold key from his pocket. "It's called a '999' key—it works like a skeleton key in that it can unlock any door in the world. I found the plans on the Internet." He places the key in my hand with instructions for use. All I need to do is shave half a millimeter off the end, insert it in a keyhole, tap the back with the handle of a screwdriver, and—voila!—access.

I ask him what is motivating his work on the ShamRocket. He immediately ticks off the three forces at work:

1. "Just to be able to do it—to build something that can fly."

2. "To get recognition for it, like, to be 'Ireland's First Rocketeer.'"

3. "And maybe to get money off of it—though I'm not sure how that would be done. That's all, then."

Has he filed any patents? No, he hadn't thought of that.

The Ulster Flying Club opened for business on June 7, 2005. The Duke of York and Prince Andrew attended the opening-day ceremonies. On one of the bloodred walls of the modest room that qualifies as the club's HQ, there is a framed letter of thanks written on behalf of the duke, signed by one Lt. Caroline Clark, Royal Navy. It reads, in part, "The Duke of York was also delighted to receive your very kind gift of the model Cessna Skyhawk II."

I have no doubt.

So that's how I ended up with a nineteen-year-old Irish theoretical-physics major with a scraggly beard and a slightly deranged look in his eye, experiencing the voluptuous panic of zero gravity. Oh, and also climbing frantically at 1.5 Gs, which is no voluptuous panic to the uninitiated—just straight panic, nothing voluptuous about it. Teeth-clenching, ass-driven-into-the-chair, quivering panic. Nausea on both counts.

But, really, it's not all awful. There's that view I mentioned earlier, for one thing—the breathtaking sweep of the purple-gray smoky jigsaw of Belfast. And the sensation is vastly different from flying in a giant commercial airplane. It is much more like an exhilarating weightless floating than the labored, rattling climb one experiences in the unpleasant tube of a 747 cabin.

I feel a sense of giddy relief when I realize Will is circling for landing. He explains a few techniques for lining up the runway and says, "Dead reckoning—I tend to use that one the most."

As we swoop in low, birdlike, I can't help but marvel aloud at what an incredible invention this is.

"Yes," Will shouts, so that I might hear through my noise-canceling headset. "Wings are wonderful things."

Though it seems as much a part of his personality as the porkpie is to Popeye Doyle, it turns out Will Breaden-Madden purchased his hat only two weeks before I met him in Niagara Falls. But the hat was effectively incorporated into his look when he noticed the reaction it generated from strangers. Some would call out, "Hey, cowboy!" and fire a finger pistol at him. Others just called him John Wayne. "That's my favorite," he tells me, as we trudge along the forever-misty streets of Belfast in search of a taxi back to campus.

Will insisted on paying for the air time—156 pounds extracted, of course, from the brown paper envelope. And it is only when we leave the Ulster Flying Club that he confesses to not really being a member of the club, exactly. He won't say how he does it, but as far as I can tell, he simply charms or scams his way in when the soaring mood strikes.

Another tattooed taxi driver slows beside us. In rainy twilight, the Queens College campus is a Goth's paradise of twisted concrete turrets and intricate Victorian facades. As we walk across a wide lawn toward his dorm room, Will lets me in on a secret prank he and a friend want to pull off: sticking a pirate's flag atop the campus's central building. Or maybe, instead, a flag featuring pi to the nth degree.

His room is barely large enough for a single bed, a desk, and a visitor. Above the desk are two plastic figurines—Einstein and Boba Fett. There are just a handful of books; Douglas Adams, the classic scientific novel *The Third Policeman,* and James Joyce among them. Tacked to a wall is the M. C. Escher drawing that's a prerequisite furnishing flair for all undergraduates, regardless of nationality, the one where the hand is trippily drawing itself.

And then there are less common dorm room images, like several sketches of human skulls. Will catches me looking at them and shouts, "I don't have a death wish or anything!"

On the windowsill sit two potted Venus flytraps and a spider under glass. Will drops a wad of notebook paper into one of the plants to demonstrate—the fangs converge. "That's all, then."

Most of the available floor space is taken up by a computer that Will's building. It's one of those prototypical, refrigerator-size jobs you see in photos from the 1960s. It's currently midproduction, so what I see is a jumble of exposed switches and wires. He picks up several tangled yellow cables, like a handful of tropical lizards. "This holds forty numbers," he tells me. To which I say nothing, staring into his arched eyebrow. "Bytes," he clarifies.

Next he shows me his homemade Theremin, which could easily be mistaken for a ham radio being attacked by a small circular saw. But there's no mistaking the creepy music it makes as pure Theremin. It's like a harp played underwater, moaning and gasping in minor keys.

"Do you know on *Star Trek,* those handheld machines that can describe the world?" he's now asking. "Well, I'm building one—it's a tricorder." By clamping sensors—tiny versions of car-battery jumper cables—to, say, a wall, the device will be able to indicate how hot the wall is. Or how high. Or thick. I have no time to contemplate this before Will is pointing out his MiG space helmet he found someplace or other, and then, last, he hands me the hacked HP-200 LX minicomputer he's rebuilt as something he calls Zico. "Go ahead, ask it a question."

I type on the miniature keypad, "What do you think of Will?" The answer streams out in blocky, pixelated letters straight from 1987: "Thinking is not possible for me. . . . I just follow my computer program. William Madden created me." I look at the machine's God, who is smiling slyly.

In addition to all this science-fair wizardry, there is one other earthly matter that Will is currently obsessed with, and that is coffee. He's only recently discovered its virtues. So much so that he's begun keeping a coffee log, wherein he meticulously chronicles, in perfectly neat penmanship, what he's consumed and where.

I flip through the pocket journal for a moment or two. "You like Starbucks' mocha."

"Oh, yes, it's quite good." Pause. "That's all, then."

We decide to hit the dorm's communal kitchen to make a pot of joe. A young female student named Allie is cooking dinner on the stove. She's cute with big teeth and furry purple slippers. Will seems to tense slightly in her presence. As Will fiddles with the beans and brewing apparatus, Allie describes the campus demographics thusly: "We have training nurses, training doctors, training lawyers," she looks at the architect of the ShamRocket, "and Will. We're not sure what Will's doing—Will, are *you* sure what you're doing?"

His eyes stay fixed on the French press. "No."

Allie: "Inventing things, I guess."

The coffee is delicious.

It's getting late—my train back to Dublin will be leaving soon. I'm tempted to hang around with Will, but I'm not sure why. I suppose I'm reluctant to surrender to the idea that the guy I felt held the most hope for helping me fulfill my quest is not going to deliver. At least not now. Maybe not ever. I badly want to believe in Will—he should totally be the Luke Skywalker of jetpacks, the young hotshot who, despite the older cynics telling him not to get cocky, trusts the Force, trusts in himself, and somehow pulls off the greatest coup of all, destroying the Death Star. Except in this case, the coup would be building a jetpack for me. This version of the story just makes sense. He's the only guy under forty I've found who is (a) really, really into this shit and (b) apparently capable of doing something about it. At this point, though, I have to admit that (b) is looking shakier by the minute. If I get in yet another taxi and wave good-bye through the back window, will I also be waving good-bye to my last best chance at solving this eighty-year riddle? I can't help but think that's the case.

But what good would hanging around Belfast do me? If Will is to be believed, and as of now, that's one mighty big if, all the goods are in a basement 147 miles south of here. And the ShamRocket is still months from completion. Besides, Will has told me he'll e-mail images of the work in progress and let me know as soon as he settles on a demonstration date. And, he says again, he'll even help pay for my way back over to see the thing fly. Or, better yet, take it for a fifteen-minute physics-defying thrill ride myself. I should just trust in the inherent goodness and honesty of man. Go back to Dublin. Go home. Hug my kids and tell them how much I missed them. Give them airplane rides and forget about all of this for a while. The ShamRocket can wait.

Which is what I do. I say good-bye to Will in front of the building he'd like to stick a pirate flag on. He presses into my hand a relay switch from his supercomputer and three packs of Beemans chewing gum. "It's what the astronauts chew," Will says. "For nausea."

Through the back window, I watch Will, standing in his Indiana Jones hat under the relentless drizzle, grow smaller as I drive away. I really have no idea if I'll ever see him after this. But one thing I do know is that if I ever see Will Breaden-Madden again, one of us will be flying a jetpack.

Epilogue

The future belongs to those who believe in the beauty
of their dreams.

—Eleanor Roosevelt

And then I'm back in Brooklyn with Catherine, the girls, all the comforts of home, and an encroaching sense of doom regarding my chances of ever flying a jetpack, let alone owning one. Moreover, any notion that our future will one day include chomping T-bone-steak meals in pills, teleporting as the spirit moves us, and getting from here to there via a fantastic flying machine appears to be a cruel joke.

There's nothing to do but cast as wide a net as possible in all directions and see if I can come up with something—*something*—containing the whiff of hope. That is to say, I begin calling everyone I can think of who might have even an inkling of a clue about any of this.

First I call the X Prize HQ in Santa Monica, the office that decided to give a million bucks to the first civilian to make it to space. A very kind if ultimately unhelpful fellow named Ian tells me that there are no plans in place to offer a similar prize to the first genius to craft a working jetpack. "We don't have anything in the works for jetpacks," Ian tells me. "But that would be cool."

Not sure exactly where to turn next, I call the World Future Society. A promising name, at least. But the future society, as envisioned by this group, will remain decidedly un-'packed. WFS rep Patrick Tucker sounds genuinely unhappy when he tells me, "I'm sorry to disappoint you, but we don't do a lot of work in jetpacks." He does say that it might be worth checking in with the local WFS branch and that perhaps I could attend a meeting (and agitate for a projetpack future, I mentally add to this thought).

So I call Harold Moore, the local chapter's president. Before dialing, I can't help it and let myself go all superstitious about Mr. Moore's name. I mean, lo these past eighty years, it has been a pair of Moores—Thomas and Wendell—who've done as much as anyone to forward jetpack technology. Could my quest meet a happy denouement with a third Moore?

As it turns out, no, it couldn't. After being called to the phone—loudly—by his wife, Harold Moore tells me the group doesn't meet much anymore and he isn't sure when they are next getting together. He has, it seems, recently had an accident. "I'm right out of the hospital," Mr. Moore says in a croaky voice. "I'm an elderly type, and I fell down and broke my noggin." His use of the word *noggin* both charms and depresses me. Rather than blasting into the future, I feel as if I'm drifting back in time, to a quainter era, by way of the world's worst time machine.

Just as I'm ready to give up for good, I reach an avuncular NASA engineer named Bruce Holmes, who has some promising news about the elusive Highway in the Sky (HITS) program. This is the program—composed of computer-driven video game–like technology that pilots can use to simplify flying small aircrafts—that companies like SoloTrek and Moller's flying-car outfit hope will one day facilitate a transition from the grounded transportation we now know to an airborne arrangement. Picture the fight scenes at the end of the first *Star Wars* movie, when Darth Vader lined up rebel X-wing pilots on a screen in his TIE fighter before lasering them down. That is approximately what a fully realized HITS system would be like, minus the lasering-down part. Holmes estimates that we are still a good ten to fifteen years away from seeing something like HITS actually in place. But he seems certain that it will eventually come to pass. Too much time and money have been spent to simply let the idea go.

The foundation for the HITS program dates back to the 1950s, when a navy engineer sought to improve on the steam gauges in plane cockpits. Back then, a pilot not wanting to land his vessel into the side of a mountain needed to quickly interpret crude heliographic images flashed on a screen on the instrument panel. The images represented what was going on outside the plane, in front of him, but provided very little scientific analysis beyond the fact that an object appeared to be in the way. Holmes tells me that the navy devised schematics of a sophisticated navigational device but didn't have the computer power necessary to get very far with it.

In the 1960s, the FAA worked up something it marvelously referred to as Project Little Guy, which would have built on the navy's plans and was designed to "keep airplanes out of trouble." But again, the technology available to realize this dream just wasn't there yet.

Still, there was clearly a demand for this sort of breakthrough in aviation. By the late 1970s, production on small, light aircraft was peaking, with nearly eighteen thousand such flying machines made in this country alone. When that trend began reversing, Holmes set out on a lecture series he called "The Role of Technology in Revitalizing the General Aviation Industry." The NASA scientist was so impassioned, in part, because he had once seen up close what this technology was capable of. Holmes once flew a jet plane utilizing an early iteration of the Highway in the Sky program. Using just a single joystick to fly through NASA airspace in Princeton, New Jersey, Holmes was able to maneuver the plane—while a computer system made the necessary adjustments to direction to help him "stay on the highway." And he wasn't alone. "We did simulacrum tests with people who had never flown in their lives, and it worked."

So wouldn't it follow, then, that one of the giant jetpacking hurdles—this whole "Well, we'd really rather not crash into one another" thing—might be overcome with a fully realized Highway in the Sky?

"If you get the Highway in the Sky program working," Holmes says, "then innovators and aircraft builders could come forth and design around that system."

Yes! So . . . basically, that means we are ten to fifteen years away from our rightful jetpacks, right?

"Is it possible?" Holmes tantalizingly and rhetorically responds to my question. "I think technologically, yes. Socially, maybe so. Economically, no question. The forces that determine how such things happen will be revolutionized in the next two decades, and all that suggests that these things are possible." Say it, Holmes, say it! "I wouldn't want to predict that there are going to be jetpacks—because in all likelihood I'd be wrong."

Sensing my clouding mood, perhaps, Bruce Holmes finally dangles this brave proclamation: "But the trend toward personalized and democratized transportation won't be held back."

I'd like to believe that when my NASA guy says this he has jetpacks in mind, but the truth is he seems much more hopeful about supersmall airplanes that can be used as taxis of the sky. And when I hear him talk about his time at the wheel of an air taxi, I have to admit, it sounds pretty good. "I find it exhilarating, absolutely delightful," Holmes says. "You fly at a low altitude—below ten thousand feet and sometimes even below five thousand—so you get to see everything. You can see kids playing in their backyards and cows grazing. It's comfortable and quiet. You feel like you are in control."

If I am completely honest with myself, I would say that this could be as good as it gets. And the part about it being comfortable, quiet, and safe is very appealing to a self-identifying wuss like me. Perhaps I've been misguided during my time hunting for a jetpack. Perhaps I should have just called up one of the many companies already building air taxis and hitched a ride. There are, too, new licensing regulations making it faster and easier than ever to be certified for flying these things. Hell, I could have been up and going, experiencing that Holmesian exhilaration, within a month of deciding to do it.

But, no. I can't kid myself. Flying a small plane is one thing—and sure, nausea over Belfast aside, I can see how pleasurable it can be. But a jetpack? To soar like a bird? That remains the ultimate fantasy.

Before hanging up, I ask him just to forget about everything he knows for a minute, to put aside all the years of studying and thinking about this stuff, and to give me his bottom line on jetpacks, that is, *will we ever have them?*

"I'm fairly agnostic when it comes to jetpacks," he begins slowly. It dawns on me that asking a scientist to ignore facts is like asking a politician to ignore lobbyists. But Holmes presses on: "Because once you can imbed controls that make it easier to fly, then the sky's the limit. Space is the limit. I know it sounds like science fiction, but that's one of the great challenges and you are in the position to take the fiction part out of it. It could get to the point that future generations just pick these things up and never even think of it as science fiction."

I really like Bruce Holmes, and he has been extremely generous with his time, but I can't help thinking that this is exactly what people were saying when Buck Rogers first rocketed off the page in the 1920s. And when the *King of the Rocketmen* matinees played to packed houses in the '40s. And when Bell took up the real-life mantle in the '50s and '60s. And Bill Suitor flew in front of billions of believers in 1984. And jetpacks became ubiquitous in pop culture from that moment on through today. Is there anything else in the world that has been so enduringly desired and never realized?

Through a friend of a friend I next found myself on the phone with Jamie Hyneman, the beret-wearing cohost of Discovery Channel's *Mythbusters*. On the show, Jamie and his crew set out to resolve myths like can a singer really break glass? Can a ninja really walk on water? So, of course, the completely captivating myth of the jetpack was right up their alley. Only they ended up building something closer to SoloTrek's ducted fan-driven Springtail and didn't get so much as a half inch off the ground. Nonetheless, if anyone out there has the time, interest, money, and smarts about this stuff and has probably at least a passing familiarity with the Highway in the Sky situation, I imagine it's Jamie.

He took time out from building a to-scale model of the *Hindenburg* that the Mythbusters would detonate later in the week to speak with me. It didn't take long to figure out that, though he is incredibly knowledgeable and generous, there was no way Jamie would be the guy to at last lead me to my rightful 'pack. For starters, he isn't even *that* into jetpacks. I was surprised to learn that growing up he wasn't the kind of kid to ogle jetpack ads in the back of *Popular Science* magazine and daydream about what it would be like to actually fly one. Rather, he is extremely grounded and pragmatic, a fan of the possible. "I don't think we are ever going to see something like George Jetson," he tells me. "It's at best a novelty. Even if somebody were to refine it to where it actually worked fairly reliably, it's still not something that is going to be mainstream; it's not going to be something that is practical to integrate into everyday life, which was the point of it: you've got it in your garage, you get into it, you commute to work with it."

He's practically describing the Toronto newspaper spread featuring Hal Graham, so many years ago. "It's not going to happen in my lifetime. Not unless there's something that I haven't seen yet that occurs that just breaks all the rules as far as being able to get people airborne without fear of falling like stones."

Right, there is that. I ask if he was afraid at all when he strapped on the would-be flying lawn mower. "I'd say the pucker factor was pretty high. I know from personal experience—you get something that has got big rotating blade–type things that are going really fast and get them near you, things happen and something goes wrong, you die. Fortunately, most of the spinning stuff was up high, and I was mainly worried about if the belts were to snap or something and come off and whip around, and I'd get flogged in the face by some kind of belt thing."

Fortunately, that didn't occur. Just before we hang up, Jamie says something that's sort of poignant and hopeful and sad all at once. "When you were a kid, maybe you felt like you could get ahold of some cardboard or something and strap it to your arms, and if you flapped hard enough you'd fly. And that's fun to think about, but very quickly if you actually start to build it realize that it ain't going to happen. But I think about that all the time. In my daydreams I'm flapping my arms real hard and I'm flying around." And just like that, I am again back in the realm of dreams.

Not long after my conversation with Jamie, a rumor began circulating through the online rocket-belt community. The chat-room whisperers were saying that Troy Widgery, the Denver entrepreneur whose Go Fast! energy-drink marketing plan includes rocket-belt demonstrations, was working on something big. Really big. Jetpack big. And this was no thirty-second air tease, the rumor went; Widgery and his technical whizzes were developing a machine that could cruise for nineteen minutes! Consecutively! At 83 mph! This could be huge.

I check the Go Fast! Web site, and there it is. They are calling it the Jet Pack T-73, in reference to the type of turbine engine it would use. It could travel twenty-seven miles. *Miles.* Not feet. And—was I really seeing this?—it would be for sale in a matter of months. That's right, for a measly two hundred thousand dollars, you, me, and anyone else who could pull together that kind of scratch would be the proud owner of a brand-new jetpack. Flight training included.

It seems Widgery, like so many others I've met this year, grew up with thoughts of James Bond dancing in his head. After launching an extreme-sports company in the early '90s, he developed a rocket belt for promo purposes. As he watched crowds from Chicago to Brazil to the Netherlands thrill to the twenty-second demos, something else began dancing in his head, too. Dollar signs. Now he apparently believes there is a real market for jetpacks. "This could be a $30 million company," he told the *Denver Business Journal* soon after posting his 'packs for sale on the Web site.

When I told Jofie all of this, he pointed out that with the T-73's specs, one could easily make it from, say, Brooklyn to Manhattan and back on a single tank of fuel. "This is the revolution," he said. And then he said it again. And again.

I called Troy Widgery and made a plan to pay a visit. Troy grew up in Denver, has a twin sister and a younger brother, and used to race midget cars when he was four years old. As a boy he watched *The Jetsons,* saw *Thunderball,* and believed he would one day take to the sky in the same way. Such boyish wonder stuck with him through his twenties, when, as a burgeoning entrepreneur, he slapped a miniature Buzz Lightyear model on the dashboard of his Range Rover, where he could watch it wriggle and shake as he snaked through downtown Denver traffic.

Troy wouldn't divulge too much info about what he was up to with his T-73, something about proprietary technology and the hope "to squeeze more power out of a small jet turbine." Hmm, could be more of a SoloTrek exoskeleton situation than an honest-to-goodness jetpack, but who am I to quibble? Not after the search I've been on. His big dream is to eventually bring the cost down so that anyone who wants to own a jetpack can. He said he is already receiving daily calls from potential buyers. "There are a bunch of Jet Pack fanatics that have contacted us," he wrote me in an e-mail. "The public's enthusiasm is very evident."

I didn't, of course, have the kind of cash on hand to snap up one of Widgery's prototypical machines, but when I told him about my quest, my year, my passion, he said I could come out and test-fly the T-73. Just as soon as it was ready, in a few months. Bill Suitor even offered to train me.

It was then that, for the first time during this whole crazy project, I became genuinely and very physically scared. My stomach went hollow. My palms moistened. I needed water. What the hell was I thinking? Was I really going to test-fly a jet-fueled metal backpack at 80 mph? The machine's kinks were still being worked out. Even my iPod didn't work too well at first; it took a few iterations to get it right. But when an iPod doesn't work, the battery can't hold a charge. When a jetpack malfunctions—suddenly I felt a tad light-headed.

As I write this, I'm still waiting for my trip to Denver. It is all, very suddenly, real. Or at least realer. I may be off soon to learn to fly a jetpack. And survive. That's the plan. Catherine, needless to say, is not a big fan of this plan. She's not amused when I remind her one night, "Hey, at least the life insurance has kicked in."

"I'm not raising our kids without a father," she says, and neither of us can believe we are unironically using dialogue straight from a bad after-school special. Our daughters are sleeping peacefully in the next room.

"You'll meet someone else," I say, and then realize that I'm basically assuming that I'm not going to survive. And that I'm easily replaceable.

"I don't want someone else. I want you not to die."

"Well, that makes two of us. And actually, I bet Troy Widgery and much of his staff feel that way, too. Can you imagine how bad that would be for his PR?"

And so if all goes well one day very soon I'm off to Colorado to finally fly a jetpack that has been tested, sure, but as far as I can tell, not
that much. If you don't hear from me again, you'll know why, but I sin-
cerely hope that's not the case. From the sounds of it, what Widgery
has come up with is not exactly the jetpack of Buck Rogers, James
Bond, George Jetson, or the Rocketeer. But it may be as close to the
dream as any of us is ever able to fly.

While I waited for the call from Troy, life moved along. One brisk fall
afternoon not long ago, Oona and I bundled up, dug her butterfly kite
out of the closet, and walked across the street to Brooklyn's biggest
park. We found a gentle slope and made a plan. I would run as fast as
I could down the slight hill and, at a certain point, throw the flimsy
plastic creation into the air while simultaneously unspooling the
thread in my other hand. Meanwhile, Oona would try to keep up.
Once I was able to steady the kite in the sky, I'd hand over controls to
the kid.

And so we ran. And then I flung. The kite flirted with the idea of
soaring, wheezed in the air making spastic fluttering noises, before
nose-diving into the brittle turf. We ran again. Fling, flutter, nosedive.
"Run, Dad, run." Fling, flutter, nosedive. We sat on a bench and called
a friend who seems like he'd be good with kite advice. He told us to
stay clear of the trees. If only that were our problem. Oona and I gave
it a couple more honest tries, but each time the butterfly crept to only
about twenty feet high before crashing back down to earth. We laid
down to rest, our breath pluming around our flushed faces as we con-
templated the sky.

"I really want the kite to fly, Dad," Oona said.

"Me, too. Maybe we need more wind."

"Maybe we do. Good idea."

Suddenly, an enormous passenger airplane rumbled through the
clouds above us, bound for Kennedy. It made flying look so easy.

"See that airplane, Dad?" Oona asked.

"Yeah, pretty amazing, huh?"

"Yeah, pretty amazing. It doesn't need more wind."

"Nope, it's doing just fine."

"Do you think it's going to California?"

"I don't think so, sweetie."

"I want to go to California someday."

"Good idea."

The plane, red lights twinkling off wings and tail in pale twilight, disappeared out of sight.

"That was beautiful, huh, Dad?"

"Very beautiful."

We scooped up the butterfly kite, walked it home, and decided to try again another day.

ACKNOWLEDGMENTS

I realize the contemporary acknowledgments page has become more like an acknowledgments novella, so I tried to keep this short. But the truth is that any book project—and especially a first book project when the writer has absolutely no idea what he is doing—is impossible to complete without the considerable kindness of friends and strangers, alike. And so, with that in mind, I'd like to thank the following people for their help and inspiration. Wendell Moore, the engine of the endeavor, for dreaming so big. Bill "Mr. Jetpack" Suitor, Hal "His Eminence" Graham, and all the generous Bell guys who shared their marvelous stories with me. Carolyn Baumet for her large-hearted retelling of family tales. Juan Lozano, Stuart Ross, Nelson Tyler, Will Breaden-Madden, Nino Amarena, Peter and Diane Ramos, and Merlin and the Stone family for graciously letting me into their homes and lives. Mark Wells for his deep knowledge of solo flight history. Joanna Ebenstein, Jofie Ferrari-Adler, Jeremy Kasten, and Aaron Ruby for their fantastic friendship and brimming enthusiasm. Heather Chaplin for the room to write and talking me down from the ledge at least three times. Peter Gijsberts, Kathleen Lennon Clough, Bill Higgins, and the excellent folks at the first ever International Rocketbelt Convention. Ben Schafer, Collin Tracy, Trent Knoss, Lissa Warren, and the entire Da Capo gang for their hard work and great instincts. Jonathan Lyons for taking a chance at lunch that day. Paul Arzt and Billy and Pinn Crawford for research, wrangling, and general radness. George Plimpton, Gay Talese and many other literary *machers* for lighting the way. The early records of Steve Martin, which shaped my worldview more than I am comfortable admitting. Matt Haber for his ongoing support and insights. Greg Mills for being so damned funny and smart. John Hyams and Nick Rhodes for their technical and creative wizardry. My brilliant and loving parents, Henri and Rose, who inspire me in ways that are both mysterious and profound. Grandma and Team Baltimore for your immense charms, incisive wits, and giant love. And, of course, the Big Three: Catherine, Oona, and Daphne—not only couldn't I have written this without you, but I wouldn't have. Thank you for this life.

INDEX